"十二五"职业教育国家规划教材
经全国职业教育教材审定委员会审定

智能制造技术专业英语

English on Intelligent Manufacturing Technology

主　编　张敬衡　王　珏
副主编　曹姗姗　刘　文
主　审　吴让大　余小燕　Edward Strzelczyk

华中科技大学出版社
中国·武汉

内容简介

为了适应智能制造技术的快速发展,满足高等职业教育改革对工学结合、任务引领的需要,高职院校教师与现代制造企业技术人员共同编写了本书。

全书内容按照智能制造类企业技术人员需要掌握的知识技术和典型工作流程,以及学生学习知识与技能的认知过程编排,主要介绍了智能制造技术、工业机器人技术、制造技术信息化、现代生产管理技术、先进制造工艺和智能制造的发展趋势等方面的内容。

本书可作为高等职业院校、高等专科院校、成人高等学校、民办高校及本科院校举办的专科层次的制造大类专业的教材,也可作为相关工程技术人员和社会从业人员的英汉对照参考书及培训用书。

图书在版编目(CIP)数据

智能制造技术专业英语/张敬衡,王珏主编. —武汉:华中科技大学出版社,2023.4
ISBN 978-7-5680-9103-9

Ⅰ.①智… Ⅱ.①张… ②王… Ⅲ.①智能制造系统-英文 Ⅳ.①TH166

中国国家版本馆 CIP 数据核字(2023)第 019616 号

智能制造技术专业英语　　　　　　　　　　　　　　张敬衡　王　珏　主编
Zhineng Zhizao Jishu Zhuanye Yingyu

策划编辑:万亚军
责任编辑:李梦阳
封面设计:原色设计
责任监印:周治超

出版发行:华中科技大学出版社(中国·武汉)　　电话:(027)81321913
　　　　　武汉市东湖新技术开发区华工科技园　　邮编:430223
录　　排:华中科技大学惠友文印中心
印　　刷:武汉市籍缘印刷厂
开　　本:787mm×1092mm　1/16
印　　张:20.25
字　　数:550千字
版　　次:2023年4月第1版第1次印刷
定　　价:59.80元

本书若有印装质量问题,请向出版社营销中心调换
全国免费服务热线:400-6679-118　竭诚为您服务
版权所有　侵权必究

前　　言

随着高新技术与传统机械行业的结合,通过企业数字化改造、智能化转型,制造业的技术不再是单纯的机械制造加工技术,而是由现代传感、信息传递、数字控制、数据采集和处理、数据库分析、网络、人工智能、现代生产管理等高新技术有机集成的智能制造技术。在我国社会经济发展持续向好和世界经济跌宕起伏缓慢复苏的背景下,国家大力推进"一带一路"建设,支持业内企业参与世界各国基础设施建设及制造业现代化建设。我国的装备制造业正从"中国制造"向"中国创造"和"中国智造"迈进,许多公司均有向海外成批输出技术与设备的业务,这些公司甚至成为全球实力最强的公司,一些企业已经担任国际行业标准制定者。随着行业的革新与发展,对制造类高素质技术技能人才的需求也随之变化,急需掌握智能制造技术、懂行业英语的专业技术技能型人才。

高等职业教育肩负着服务区域经济,为国有、民营乃至外向型企业提供高素质人才的战略任务。因此,人才培养目标中增加掌握外语及国际行业标准的要求就显得尤其重要。为使毕业生适应我国工业 4.0 发展需要,使他们既能承担智能制造企业一线新岗位的工作,又能胜任外向型制造企业的岗位工作,并能随企业海外项目走向世界,就必须加强专业外语的教育教学。

智能制造技术专业英语课程是制造大类专业通用的一门专业课,课程目标是帮助学生建立专业英语意识,使学生具备掌握基本专业词汇、看懂常用工具资料、运用英语现场工作的能力。

智能制造技术专业英语课程的开设急需相关教材。本书是在原有的"十二五"职业教育国家规划教材《现代制造技术专业英语(第三版)》的基础上,借鉴当前国内外智能制造技术实践成果和发展趋势,参考大量智能制造技术英语原版资料、产品说明书及操作维护手册等专业资料,由高职院校教师与现代制造企业技术人员共同编写的。本书旨在帮助相关专业的学生在学习英语的同时,进一步了解和掌握智能制造知识和技术,为我国制造类企业普及和推广工业 4.0 及智能制造,培养懂技术、会英语的应用型人才。

全书主要包括智能制造技术、工业机器人技术、制造技术信息化、现代生产管理技术、先进制造工艺和智能制造的发展趋势等方面的内容,共分为 6 个单元,每个单元 5 篇课文,共 30 篇。同时,每个单元附有与课文内容相关的 4 篇拓展阅读材料,共 24 篇。本书还附有词汇表、缩略语词汇表。考虑到学习的连贯性和持续性,全书参考学时数为 80 个学时,分两个学期完成。学习重点放在阅读理解、专业词汇积累、书面英译汉和知识技术拓展上。

本书可作为高等职业院校、高等专科院校、职业本科、成人高等学校、民办高校及五年制高等职业院校制造类专业(如教育部高等职业教育专业目录中的工业机器人技术等自动化类专业、智能制造装备技术等机电设备类专业、数控技术等机械设计制造类专业、汽车制造与试验技术等汽车制造类专业)的教学用书,也可作为相关工程技术人员和社会从业人员的英汉对照参考书及培训用书。

本书体现了人才培养的层次性、知识结构的交融性和教学内容的实践性;作为外语教学材料,注重知识面的拓展;将教学内容按知识构成与实践操作需要进行整合,努力填补高职高专教育教学中智能制造技术英语教材的空白。

(1)全部章节采用"单元导言"导入式教学,部分课文采用案例式教学方法,力求做到学习内容联系工作实际,知识的宽度和深度循序渐进。倡导自主学习,本书专业单词注有音标,使学生可以通过开口练习来尽快熟悉专业英语表达,并积累一定数量的专业词汇;配备课文注解、页边提示和小贴士,方便学生阅读;每篇课文均安排了练习,方便学生及时巩固所学;书末词汇表提供课文回溯,方便学生在语境中掌握生词。课文图文并茂,界面友好。

(2)以近年来推广的智能制造技术为主,兼顾制造专业大类以及相关的新科技、新技术,本书适当介绍了互联网、云计算、物联网、大数据、射频识别技术、数字孪生技术、增强现实技术、机器视觉、人工智能等现代科学技术及其在制造业中的应用,对培养学生学习兴趣和拓宽其知识面起到了积极作用。

(3)本书所选英语资料全部来自原版资料或我国外向型企业产品说明书等资料,用词、句型、语法结构规范,有利于培养学生使用纯正英语的习惯,避免"自创"英语,培养学生使用智能制造技术原版资料的能力。

(4)根据语言学习的特点,激发学生学习英语的潜能,书中以二维码形式链接课文录音供学生练习口语和正音。为学生能试着开口说专业的话营造语言环境,提倡专业英语教学兼顾听、说能力的培养,帮助学生加强对课文内容的记忆和理解,并为对外技术交流打下一定的基础,使学生更有学成的自豪感。

(5)为了方便教师的教学,本书配有教学PPT模板及练习答案(联系邮箱:jixie_hustp@163.com)。每个单元附有与单元课文相近的阅读材料,可供教师做测试选题;也可让学生课外练习阅读、翻译,巩固和拓宽学习内容。为了便于学生自学,所有课文和阅读材料均附有参考译文,其以网络资源形式呈现。

本书由张敬衡、王珏担任主编,曹姗姗、刘文担任副主编。来自企业的周理、宁柯、孙海亮、石义淮、孙立勇等技术专家和来自高职院校的雷黎明、陈江进、龚五堂、陈璇、刘芳、韩玮等参与编写。第一单元由张敬衡、周理、雷黎明、刘文编写;第二单元由石义淮、张敬衡、韩玮、曹姗姗编写;第三单元由张敬衡、孙海亮、韩玮、刘文编写;第四单元由王珏、宁柯、曹姗姗、刘芳编写;第五单元由孙立勇、王珏、陈江进、陈璇编写;第六单元由张敬衡、龚五堂、陈璇、刘芳编写。参考译文由王珏、张敬衡、刘文整理;词汇表由曹姗姗、刘文整理;缩略语词汇表由刘文、曹姗姗整理。张敬衡、王珏对全书进行内容选题、修改和统稿。

本书由企业专家吴让大先生、高校教授余小燕老师、澳大利亚机电专家 Edward Strzelczyk 先生担任主审。

由于编者水平有限,加上形势的发展也在不断提出新的要求,书中难免有疏忽和错误之处,敬请读者批评指正。

编 者
2022年7月于武汉

Contents

Unit I Intelligent Manufacturing Technology ········ 1
- Lesson 1 The Concept of Intelligent Manufacturing Technology ········ 1
- Lesson 2 Innovation and Characteristics of Industry 4.0 ········ 6
- Lesson 3 The Application of Artificial Intelligence in Manufacturing Industry ········ 14
- Lesson 4 Application of RFID Technology in Intelligent Manufacturing ········ 21
- Lesson 5 Intelligent Robot ········ 27
- Reading Material 1 Artificial Intelligence ········ 32
- Reading Material 2 Expert System ········ 35
- Reading Material 3 Cloud Manufacturing ········ 38
- Reading Material 4 Computer Vision ········ 41

Unit II Industrial Robot Technology ········ 44
- Lesson 6 Industrial Robot ········ 44
- Lesson 7 The Maintenance of the Manipulator ········ 51
- Lesson 8 The Maintenance of the Manipulator Control Cabinet ········ 58
- Lesson 9 Autonomous Mobile Robotics ········ 66
- Lesson 10 Basic Operation Training Tasks of Industrial Robots ········ 71
- Reading Material 5 Installation of the Manipulator ········ 75
- Reading Material 6 Robot Parts List ········ 78
- Reading Material 7 Emergency Safety Information ········ 82
- Reading Material 8 The Robot Palletizer ········ 87

Unit III Manufacturing Technology Informatization ········ 91
- Lesson 11 Wireless Connectivity for Industries ········ 91
- Lesson 12 Application of Digital Twin Technology in Manufacturing Industry ········ 98
- Lesson 13 Application of AR Technology in Manufacturing Industry ········ 104
- Lesson 14 Automated Guided Vehicles ········ 111
- Lesson 15 Teach Pendant Manual for Controller-B ········ 117
- Reading Material 9 The Internet of Things ········ 123
- Reading Material 10 Multimedia Technology ········ 126
- Reading Material 11 Global Positioning System (GPS) ········ 130
- Reading Material 12 Virtual Instruments ········ 133

Unit IV Modern Production Management Technology ········ 137
- Lesson 16 Asset Status Monitoring ········ 137
- Lesson 17 Manufacturing Execution System ········ 143
- Lesson 18 Computer-aided Process Planning ········ 150

Lesson 19	Automatic Detection of Weld Defects	155
Lesson 20	Intelligent Manufacturing Control MES Software	162
Reading Material 13	Product Data Management	167
Reading Material 14	Manufacturing Resource Planning	170
Reading Material 15	Quality Control	174
Reading Material 16	Enterprise Resource Planning	178

Unit Ⅴ Advanced Manufacturing Process — 181

Lesson 21	Micromachining	181
Lesson 22	Laser Processing	188
Lesson 23	EDM Machine	195
Lesson 24	Additive Manufacturing	200
Lesson 25	Manual Programming Training of CNC Machining Center	206
Reading Material 17	3D Printing Technology	214
Reading Material 18	Nanotechnology	217
Reading Material 19	High-speed Machining	221
Reading Material 20	Flexible Manufacturing System	224

Unit Ⅵ Development Trend of Intelligent Manufacturing — 228

Lesson 26	Agile Manufacturing	228
Lesson 27	Green Manufacturing	234
Lesson 28	Lean Production	239
Lesson 29	Computer-integrated Manufacturing	244
Lesson 30	Collaborative Robots	249
Reading Material 21	Machine Learning	256
Reading Material 22	Digital Image Processing	258
Reading Material 23	The Latest Intelligent Manufacturing Technology	260
Reading Material 24	Intelligent Manufacturing Led by New Generation of AI Technology	264

Appendix Ⅰ Vocabulary — 267

Appendix Ⅱ Abbreviations — 310

References — 313

目　录

第一单元　智能制造技术 ... 1
第 1 课　智能制造技术的概念 ... 1
第 2 课　工业 4.0 的创新与特点 ... 6
第 3 课　人工智能在制造业中的应用 ... 14
第 4 课　RFID 技术在智能制造中的应用 ... 21
第 5 课　智能机器人 ... 27
阅读材料 1　人工智能 ... 32
阅读材料 2　专家系统 ... 35
阅读材料 3　云制造 ... 38
阅读材料 4　计算机视觉 ... 41

第二单元　工业机器人技术 ... 44
第 6 课　工业机器人 ... 44
第 7 课　机械手的保养 ... 51
第 8 课　机械手控制柜的维护 ... 58
第 9 课　自主移动机器人 ... 66
第 10 课　工业机器人的基本操作实训任务 ... 71
阅读材料 5　机械手的安装 ... 75
阅读材料 6　机器人零件清单 ... 78
阅读材料 7　紧急安全信息 ... 82
阅读材料 8　机器人码垛工 ... 87

第三单元　制造技术信息化 ... 91
第 11 课　工业无线连接 ... 91
第 12 课　数字孪生技术在制造业中的应用 ... 98
第 13 课　增强现实技术在制造业中的应用 ... 104
第 14 课　自动导引车 ... 111
第 15 课　用于 B 型控制器的示教器使用手册 ... 117
阅读材料 9　物联网 ... 123
阅读材料 10　多媒体技术 ... 126
阅读材料 11　全球定位系统 ... 130
阅读材料 12　虚拟仪器 ... 133

第四单元　现代生产管理技术 ... 137
第 16 课　资产监控 ... 137
第 17 课　制造执行系统 ... 143
第 18 课　计算机辅助工艺设计 ... 150

第 19 课	焊接缺陷自动检测	155
第 20 课	智能制造管控 MES 软件	162
阅读材料 13	产品数据管理	167
阅读资料 14	制造资源计划	170
阅读材料 15	质量控制	174
阅读材料 16	企业资源计划	178

第五单元　先进制造工艺　　181

第 21 课	微细加工	181
第 22 课	激光加工	188
第 23 课	电火花加工机床	195
第 24 课	增材制造	200
第 25 课	数控加工中心的手工编程实训	206
阅读材料 17	3D 打印技术	214
阅读材料 18	纳米技术	217
阅读材料 19	高速加工	221
阅读材料 20	柔性制造系统	224

第六单元　智能制造的发展趋势　　228

第 26 课	敏捷制造	228
第 27 课	绿色制造	234
第 28 课	精益生产	239
第 29 课	计算机集成制造	244
第 30 课	协作机器人	249
阅读材料 21	机器学习	256
阅读材料 22	数字图像处理	258
阅读材料 23	最新智能制造技术	260
阅读材料 24	新一代人工智能技术引领下的智能制造	264

附录一　词汇表　　267

附录二　缩略语词汇表　　310

参考文献　　313

Unit Ⅰ　Intelligent Manufacturing Technology

Introduction to the Unit 单元导言

> 本单元的任务是在英语环境下,了解智能制造技术概念以及当前制造业创新热点内容,例如:工业4.0的创新与特点,人工智能在制造业中的应用,RFID技术在智能制造中的应用,智能机器人等。本单元的目的是使学生在学习英语的过程中,复习和新增智能制造类专业相关知识,巩固和强化已有的英语知识,并将语言与专业技术有机结合,在理解的基础上掌握制造业英语常用表达方式、典型句型和专业术语,建立专业英语意识,为深入学习后续课程打下基础。

Lesson 1　The Concept of Intelligent Manufacturing Technology

　　Intelligent manufacturing (IM) system is a human-machine integrated intelligent system, which is composed of[①] intelligent machines and human experts. It can carry out intelligent activities in the manufacturing process, such as analysis, reasoning, judgment, conception, and decision-making. Through the cooperation of humans and intelligent machines, we will expand, extend and partially replace the brain work of human experts in the manufacturing process. It updates the concept of manufacturing automation, which is flexible, intelligent and highly integrated.

　　Intelligent manufacturing comes from research on artificial intelligence, including intelligent manufacturing technology (as shown in Fig. 1-1) and intelligent manufacturing systems. Intelligent manufacturing systems can not only constantly enrich the knowledge base in practice with the function of independent learning but also be able to collect and understand environmental information and its own information. Besides, it can judge and plan its own behavior.

　　Artificial intelligence technology is widely used in almost every link

Margin Note

integrated
intelligent system
一体化智能系统

artificial
intelligence
人工智能

of the manufacturing process. Expert system technology can be used in engineering design, process design, production scheduling, fault diagnosis, and so on. In addition②, it can apply advanced computer intelligence methods such as neural networks and fuzzy control technology to product formulation, and production scheduling to realize intelligent manufacturing processes. In order to understand intelligent manufacturing, two important concepts need to be established, namely agent and holonic system.

production scheduling
生产调度

fault diagnosis
故障诊断

fuzzy control technology
模糊控制技术

holonic system
整子系统

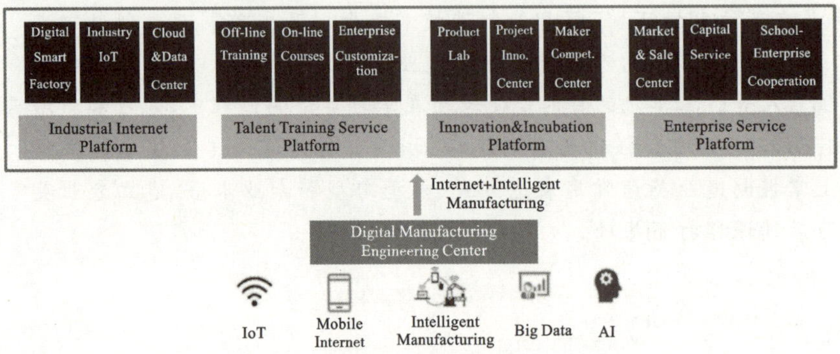

Fig. 1-1 Intelligent Manufacturing Technology

In IT, software or hardware entities capable of③ autonomous activities are called agents. With the wide application of artificial intelligence and computer technology in the manufacturing industry, multi-agent system (MAS) (as shown in Fig. 1-2) technology provides an intelligent method to solve the coordination and cooperation between multiple fields in product design, production, and manufacturing, and even the whole life cycle of products. It also provides the method for system integration and parallel design. Thus, intelligent manufacturing is realized.

multi-agent system (MAS)
多智能体系统

Holon is the smallest component of many different kinds that constitute the holonic system. Holon is autonomous (each holon operates according to the plan and responds autonomously to task changes and accidents with controllable behavior), cooperative (each holon can request other holons to perform a certain operation, and provide services as well), and intelligent (intelligence of reasoning, judgment, etc.). It is similar to the concept of an agent. Therefore, the holonic system has two characteristics: agility and flexibility. It has a strong self-organization ability with agility to build new systems quickly and reliably and flexibility to adjust manufacturing to meet the requirements of rapid market changes.

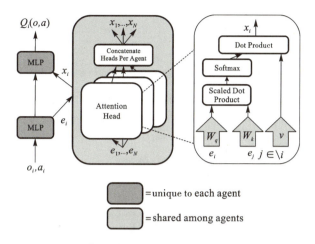

Fig. 1-2　Multi-agent　System

According to the basic idea of distributed integration, the basic principle of intelligent manufacturing is to apply multi-agent system technology to realize the flexible and intelligent integration of manufacturing units and network-based manufacturing systems in the distributed manufacturing network environment. On the basis of[④] the implementation of each local area of an intelligent manufacturing system, the intelligent manufacturing system is realized in the global manufacturing network environment based on the Internet.

distributed integration 分布式集成

 Tips

1. Fuzzy Control Technology 模糊控制技术

模糊控制技术是近代控制理论中的一种高级策略和新颖技术。模糊控制技术基于模糊数学理论,通过模拟人的近似推理和综合决策过程,使控制算法的可控性、适应性和合理性得以提高,是智能控制技术的一个重要分支。

2. MAS:Multi-agent System 多智能体系统

多智能体系统技术是一种全新的分布式计算技术,自20世纪70年代出现以来得到迅速发展,已经成为一种进行复杂系统分析与模拟的方法与工具。一个多智能体系统,是由在一个环境中交互的多个智能体组成的计算系统。多智能体系统也能被用来解决分离的智能体以及单层系统难以解决的问题。智能可以由一些方法、函数、过程、搜索算法或加强学习来实现。

New Words and Phrases

integrated [ˈɪntɪɡreɪtɪd] adj.综合的,完整统一的,各部分密切协调的
conception [kənˈsepʃn] n.概念,观念,构想,设想
partially [ˈpɑːʃəli] adv.部分地,不完全地
replace [rɪˈpleɪs] v.代替,取代
flexible [ˈfleksəbl] adj.灵活的,易变通的,适应性强的
constantly [ˈkɒnstəntli] adv.始终,一直,重复不断地
fault diagnosis 故障诊断
digital [ˈdɪdʒɪtl] adj.数字的,数码的,数字信息系统的
entity [ˈentəti] n.实体
autonomous [ɔːˈtɒnəməs] adj.自主的,有自主权的
application [ˌæplɪˈkeɪʃn] n.(尤指理论、发现等的)应用,运用
coordination [kəʊˌɔːdɪˈneɪʃn] n.协作,协调,配合
multiple [ˈmʌltɪpl] adj.数量多的,多种多样的
parallel [ˈpærəlel] adj.并行的,平行的
component [kəmˈpəʊnənt] n.组成部分,成分,部件
constitute [ˈkɒnstɪtjuːt] v.组成,构成
agility [əˈdʒɪlɪti] n.敏捷性,灵敏性
concatenate [kɒnˈkæt(ə)ˌneɪt] vt.连接,拼接 adj.连锁的
implementation [ˌɪmplɪmɛnˈteɪʃən] n.履行,实施
integrated intelligent system 一体化智能系统
artificial [ˌɑːtɪˈfɪʃl] adj.人工的,人造的
artificial intelligence 人工智能
scheduling [ˈʃedʒuːəlɪŋ] v.调度,制定时间表 n.行程安排
production scheduling 生产调度
fuzzy control technology 模糊控制技术
multi-agent system 多智能体系统
holonic [həʊˈlɒnɪk] adj.整子的,子整体的,合子
holonic system 整子系统
distributed [dɪˈstrɪbjuːtɪd] adj.分布式的
distributed integration 分布式集成

 Notes

1. (be) composed of 由……组成

例句:Intelligent manufacturing (IM) system is a human-machine integrated intelligent system composed of intelligent machines and human experts.

智能制造是一种由智能机器和人类专家共同组成的人机一体化智能系统。

例句:Every substance, no matter what it is, is composed of very small particles called molecules.

所有物质,不论它是什么,都是由一些被称为分子的很小的粒子构成的。

2. in addition (to) 此外,另外,加之,除……之外

例句:In addition, it can apply advanced computer intelligence methods such as neural networks and fuzzy control technology to product formulation, and production scheduling to realize intelligent manufacturing processes.

另外,它也可以将神经网络和模糊控制技术等先进的计算机智能方法应用于产品规划、生产调度,实现制造过程智能化。

例句:In addition to preparing for the interview, you'll also learn whether or not the company and its culture are a right fit for you.

除了能为面试做准备,你还能看看企业及其文化对你来说是否是合适的。

3. (be) capable of 有……能力的,可……的

例句:In IT, software or hardware entities capable of autonomous activities are called agents.

在信息技术(IT)领域,能够自主运行的软件或者硬件实体称为智能体。

例句:The idea that software is capable of any task is broadly true in theory.

软件能够处理任何任务的观念从理论上说基本是正确的。

4. on the basis of 基于,基础上

例句:On the basis of the implementation of each local area of an intelligent manufacturing system, the intelligent manufacturing system is realized in the global manufacturing network environment based on the Internet.

在实现智能制造系统的各个局域的基础上,实现基于互联网的全球制造网络环境下的智能制造系统。

例句:A good parent-children relationship should be set up on the basis of mutual understanding and respect.

良好的亲子关系应该建立在相互理解和尊重的基础上。

Exercises

Ⅰ.Write True or False beside the following statements about the text.

1. _____ Intelligent manufacturing can carry out intelligent activities in the manufacturing process.

2. _____ Intelligent manufacturing systems can not only enrich the knowledge base in practice, but also be able to collect and understand environmental information and its

own information, judge and plan its own behavior.

3. _____ To understand intelligent manufacturing, you need to understand two important concepts, namely agent and holonic system.

4. _____ In IT, software or hardware entities capable of autonomous activities are called agents.

5. _____ Holon is similar to the concept of an agent.

Ⅱ. Answer the following questions in English according to the text.

1. What is intelligent manufacturing?
2. What can we do through the cooperation of humans and intelligent machines?
3. What is a holon? Why do we say holon is cooperative?
4. What are the characteristics of the holonic system?

Ⅲ. Read the text again and fill in the blanks in the following sentences orally.

1. Intelligent manufacturing (IM) system is a human-machine integrated intelligent system composed of _____ _____ and _____ _____.

2. Intelligent manufacturing systems can not only _____ _____ the knowledge base in practice with the _____ of independent learning but also be able to _____ and _____ environmental information and its own information.

3. Multi-agent system technology provides an intelligent method to solve the _____ and _____ between multiple fields.

4. Each holon operates according to the plan and responds _____ to task changes and accidents with _____ _____.

5. _____ _____ the basic idea of distributed integration, the basic principle of intelligent manufacturing is to _____ multi-agent system technology to _____ the flexible and intelligent integration of manufacturing units and network-based manufacturing systems in the distributed manufacturing network environment.

Ⅳ. Translation.

Holon is the smallest component of many different kinds that constitute the holonic system. Holon is autonomous, cooperative, and intelligent. It is similar to the concept of an agent. Therefore, the holonic system has two characteristics: agility and flexibility. It has a strong self-organization ability with agility to build new systems quickly and reliably and flexibility to adjust manufacturing to meet the requirements of rapid market changes.

Lesson 2　Innovation and Characteristics of Industry 4.0

The four industrial revolutions significantly impacted the　　Margin Note manufacturing process of the time (as shown in Fig. 2-1). The 1st industrial revolution introduced mechanical production facilities supported by water and steam power, marking the beginning of the age of mechanization. Mass production assembly line fuels the 2nd industrial revolution with the help of electrical power. Combining electronics, computers, IT, and robotics allows for further automation of the production process in the 3rd industrial revolution. The connected enterprise leads to the 4th industrial revolution, connecting production facilities with the internet of things through the smart factory, autonomous systems, and machine learning (as shown in Fig. 2-2).

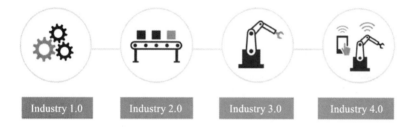

Fig. 2-1 The Industrial Revolutions

Fig. 2-2 Internet of Things (IoT)

internet of things (IoT)
物联网

In essence, Industry 4.0 is the trend towards automation and data exchange in manufacturing technologies and processes which include cyber-physical systems (CPSs), internet of things, industrial internet of things (as shown in Fig. 2-3), cloud computing, cognitive computing, and artificial intelligence (AI).

cyber-physical system
信息物理系统

Industry 4.0 has four key innovations:

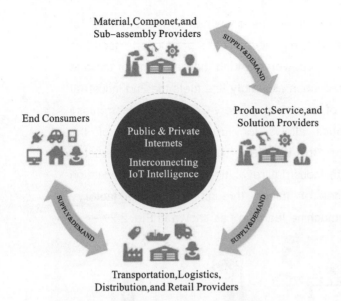

Fig. 2-3 Industrial Internet of Things (IIoT)

(1) Interconnection — the ability of machines, devices, sensors, and people to connect and communicate with each other via the internet of things, or the internet of people (IoP).

(2) Information transparency — the transparency afforded by Industry 4.0 technology provides operators with comprehensive information to make decisions. Inter-connectivity allows operators to collect immense amounts of① data and information from all points in the manufacturing process, thus identifying key areas that can benefit from② the improvement to increase functionality.

(3) Technical assistance — the technological facility of systems to assist humans in decision-making and problem-solving, and help humans deal with③ difficult or unsafe tasks.

(4) Decentralized decisions — the ability of cyber-physical systems to make decisions on their own and to perform their tasks as autonomously as possible. Only in the case of exceptions, interference, or conflicting goals, are tasks delegated to a higher level.

Intelligent manufacturing plays an important role in Industry 4.0 (as shown in Fig. 2-4) which is used in many industries, such as the automotive industry, logistics industry, construction industry, and electric power industry.

Industry 4.0 has the following four characteristics:

(1) Velocity — the exponential speed at which incumbent industries are affected and displaced.

(2) Scope and systems impact — the large number of sectors and firms that are affected.

internet of people (IoP)
人联网

inter-connectivity
互联性

intelligent manufacturing
智能制造

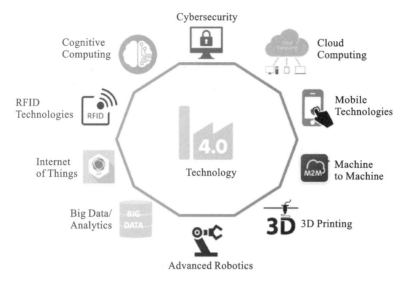

Fig. 2-4　Industry 4.0

(3) A paradigm shift in technology policy — new policies designed for this new approach (Industry 4.0) have emerged. Many countries have officially recognized Industry 4.0 in their innovation policies.

(4) Intelligent factory — the intelligent factory is based on cyber-physical systems that communicate with each other using the internet of things and services. An important part of this process is the exchange of data between the product and the production line, which enables a much more efficient connection of the supply chain and better organization within any production environment.

supply chain
供应链

Using cyber-physical systems that monitor physical processes, a virtual copy of the physical world can be designed. Cyber-physical systems are capable of making decentralized decisions independently, reaching a high degree of autonomy. Virtual simulations require data, which smart IoT sensors deliver to them in real-time. With the support of technologies such as the internet of things, AI, cloud computing, and digital twinning, continuous integration of the latest information can provide the latest view of products and processes, enabling enterprises to proactively, predictably, and cost-effectively prevent failures and find solutions in product development, helping to formulate effective business strategies. In production practices, this means reducing downtime, reducing production and maintenance costs, enhancing customer service capabilities, and reducing time-to-market.

virtual copy
虚拟副本

digital twin
数字孪生

We are in the age of Industry 4.0. Evolution and advancements in information and communication technology, sensors, big data, internet of things, 3D printing, cloud computing, robots, and mobile internet are some of the key technology areas that will digitize the value chains in various industries. These technologies can be summarized into four major components: cyber-physical systems, internet of things, on-demand availability of computer system resources, and cognitive computing. Besides, they are defined as Industry 4.0 or smart factories. Therefore, we can say that it is the application of various new technologies in Industry 4.0 to the manufacturing industry that promotes the rapid development of "intelligent factories".

Tips

1. CPS: Cyber-physical System 信息物理系统

信息物理系统是一个综合计算、网络和物理环境的多维复杂系统,通过3C (computation, communication, control) 技术的有机融合与深度协作,实现大型工程系统的实时感知、动态控制和信息服务。CPS 实现计算、通信与物理系统的一体化设计,通过人机交互接口实现和物理进程的交互,使用网络化空间以远程、可靠、实时、安全、协作的方式操控物理实体。

2. Cognitive Computing 认知计算

认知计算是认知科学的核心技术子领域之一,是人工智能的重要组成部分,用来模拟人脑认知过程。它包含信息分析、自然语言处理和机器学习领域的大量技术创新,能够助力决策者从大量非结构化数据中洞察规律。认知系统能够以更加自然的方式与人类交互;可获取海量的不同类型的数据,根据信息进行推论;从自身与数据、与人类的交互中学习。

3. IIoT: Industrial Internet of Things 工业物联网

工业物联网是指将具有感知、监控能力的各类采集、控制传感器或控制器,以及移动通信、智能分析等技术不断融入工业生产过程各个环节,从而大幅提高制造效率,改善产品质量,降低产品成本,减少资源消耗,最终实现将传统工业提升到智能化的新阶段。从应用形式上看,工业物联网的应用具有实时性、自动化、嵌入式(软件)、安全性和信息互通互联性等特点。

4. IoP: Internet of People 人联网

人联网,并不是一个与物联网相对立的概念,而是一个包容物联网、传统讯息内容互联网,以及服务互联网的综合构架。它是指以人为核心,以移动互联网为主载,强调人的实时、互动、体验,融合虚拟世界与实体世界,提供综合一体的网络应用解决方案。

New Words and Phrases

significantly [sɪɡˈnɪfɪkəntli] adv. 显著地,明显地
facility [fəˈsɪləti] n. 设施,设备
electronics [ɪˌlekˈtrɒnɪks] n. 电子技术,电子学
innovation [ˌɪnəˈveɪʃn] n. (新事物、思想或方法的)创造,创新,改革
interconnection [ˌɪntəkəˈnekʃn] n. 互联互通,紧密联系,关联
device [dɪˈvaɪs] n. 装置,仪器,器具,设备
sensor [ˈsensə(r)] n. 传感器,敏感元件,探测设备
transparency [trænsˈpærənsi] n. 透明度,清晰度
comprehensive [ˌkɒmprɪˈhensɪv] adj. 综合的,全面的
identify [aɪˈdentɪfaɪ] v. 确认,认出,鉴定
functionality [ˌfʌŋkʃəˈnæləti] n. 功能性
decentralize [ˌdiːˈsentrəlaɪz] v. 分散,分权,使(业务)分散,疏散(人口)
interference [ˌɪntəˈfɪərəns] n. 干涉,干预,介入
conflict [ˈkɒnflɪkt] v. 冲突,抵触
delegate [ˈdelɪɡeɪt] v. 授(权),把(职责、责任等)委托(给)
velocity [vəˈlɒsəti] n. (沿某一方向的)速度
exponential [ˌekspəˈnenʃl] n. 指数函数 adj. 指数的,迅猛的,呈几何级数的
automotive [ˌɔːtəˈməʊtɪv] adj. 汽车的,自动的
logistics [ləˈdʒɪstɪks] n. 物流,后勤学,运筹学,统筹安排
incumbent [ɪnˈkʌmbənt] adj. 在职的,现任的
displace [dɪsˈpleɪs] v. 取代,替代,置换
paradigm [ˈpærədaɪm] n. 样式,典范,范例
monitor [ˈmɒnɪtə(r)] v. 监视,检查,跟踪调查
simulation [ˌsɪmjuˈleɪʃn] n. 模仿,仿真
virtual simulation 虚拟仿真
proactively [ˌprəʊˈæktɪvli] adv. 积极主动地,主动出击地,先发制人地
predictably [prɪˈdɪktəbli] adv. 可预言地,可预测地,可预料地
cost-effectively 有成本效益地,划算地
downtime [ˈdaʊntaɪm] n. (尤指计算机的)停机时间,停止运行时间
maintenance [ˈmeɪntənəns] n. 维护,保养
digitize [ˈdɪdʒɪtaɪz] v. 使数字化
time-to-market 上市时间
on-demand 按需的
cyber-physical system 信息物理系统
internet of things 物联网

internet of people 人联网
inter-connectivity 互联性
manufacturing [ˌmænjuˈfæktʃərɪŋ] n. 制造
intelligent manufacturing 智能制造
supply chain 供应链
virtual [ˈvɜːtʃuəl] adj. 虚拟的，模拟的
virtual copy 虚拟副本
digital twin 数字孪生

 Notes

1. amounts of 大量的

例句：Inter-connectivity allows operators to collect immense <u>amounts of</u> data and information from all points in the manufacturing process...

互联互通使运营商能够从制造过程中的各个点收集大量数据和信息……

例句：The server is designed to store huge <u>amounts of</u> data.

该服务器是为存储大量数据设计的。

2. benefit from 从……中获益

例句：Inter-connectivity allows operators to collect immense amounts of data and information from all points in the manufacturing process, thus identifying key areas that can <u>benefit from</u> the improvement to increase functionality.

互联互通使运营商能够从制造过程中的各个点收集大量数据和信息，从而确定可以通过改进来<u>获益</u>的关键领域，以增进功能。

例句：The sensor can be used for a variety of applications, which <u>benefits from</u> the small dimensions and high sensitivity.

<u>得益于</u>其小体积和高敏感性，传感器能适用于各种各样领域。

3. deal with 处理，应付，与……打交道

例句：Technical assistance — the technological facility of systems to assist humans in decision-making and problem-solving, and help humans <u>deal with</u> difficult or unsafe tasks.

技术援助——帮助人类决策和解决问题的，以及帮助人类完成困难或不安全任务的系统的技术设施。

例句：We must take positive steps to <u>deal with</u> the problem.

我们必须采取积极措施来<u>处理</u>这个问题。

Exercises

Ⅰ. Write True or False beside the following statements about the text.

1. _____ There are four key innovations mentioned in the text.

2. _____ Transparency is afforded by Industry 4.0 technology, which provides operators with comprehensive information to make decisions.

3. _____ Cyber-physical systems should work with humans to make decisions and perform their tasks.

4. _____ Smart IoT sensors deliver the data that is required by virtual simulations.

5. _____ Without the application of new technologies in Industry 4.0 to the manufacturing industry, "intelligent factories" would not be developed rapidly.

Ⅱ. Answer the following questions in English according to the text.

1. What are the four key innovations of Industry 4.0?

2. What are the benefits of information transparency in the manufacturing process?

3. Why is the exchange of data between the product and the production line so important?

Ⅲ. Read the text again and fill in the blanks in the following sentences orally.

1. Interconnection refers to _____ _____ of machines, devices, sensors, and people to _____ and _____ with each other _____ the internet of things, or the internet of people (IoP).

2. The transparency _____ by Industry 4.0 technology provides operators _____ comprehensive information to _____ _____.

3. The intelligent factory _____ _____ _____ cyber-physical systems that communicate with each other _____ the internet of things and services.

4. The exchange of data between the product and the production line _____ a much more efficient connection of the supply chain and better organization within any production environment.

5. With the support of technologies such as the internet of things, AI, cloud computing, and digital twinning, continuous _____ of the latest information can provide the latest view of products and processes, helping enterprises _____ _____ effective business strategies.

Ⅳ. Translation.

The intelligent factory is based on cyber-physical systems that communicate with each other using the internet of things and services. An important part of this process is the exchange of data between the product and the production line, which enables a much more efficient connection of the supply chain and better organization within any production environment.

Lesson 3 The Application of Artificial Intelligence in Manufacturing Industry

Artificial intelligence (AI) in manufacturing is the intelligence in which machines perform human-like tasks autonomously (as shown in Fig. 3-1) and respond to events inside and around the machine. Intelligent machine tools can automatically detect tool wear or faults, and even predict faults that will occur, react and solve them. Manufacturing AI enables more accurate process design, process diagnosis, and troubleshooting by using digital twin technology. Digital twin technology is not only a computer-aided design (CAD) model but also an accurate virtual copy of the actual parts, machine tools, or workpiece being processed. In addition, it is an accurate digital representation of the processing and whether parts are qualified or not, presenting images and solving the problems.

Margin Note

machine tool
机床

process diagnosis
过程诊断

computer-aided design
计算机辅助设计

Fig. 3-1 Machines Perform Human-like Tasks Autonomously

Artificial intelligence technology has given rise to[①] "boxed factories" in manufacturing (as shown in Fig. 3-2). When a machine is delivered with an AI device, the manufacturer packages the end-to-end workflow and provides the user with installation instructions, knowledge references, analytical methods for sensor detection operations and machine maintenance, and unsupervised models. The user is trained to create a boxed factory system by using an unsupervised model to look for anomalies or errors and be able to compare them with the normal feedback patterns of the sensors. Such a system would allow users to

end-to-end workflow
端到端工作流程

sensor detection
传感器检测

check parts produced today, compare them with those produced yesterday, and analyze nondestructive testing for each process along the production line. This will help users understand exactly what parameters are used to manufacture these parts and then trace defects from sensor data. The idea behind the "boxed factory" process is to load raw materials at one end and take finished parts out at the other, leaving technicians to maintain the system.

nondestructive testing 无损检测

raw material 原材料,原料

finished part 成品零件

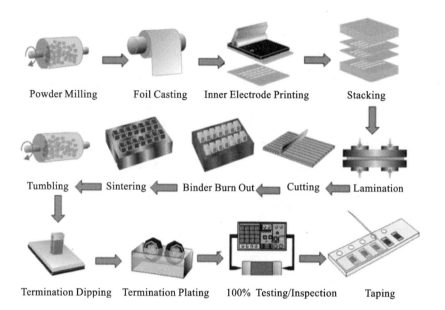

Fig. 3-2　One of the "Boxed Factory"

The realization of artificial intelligence comes largely from machine learning, neural networks, deep learning, and other self-organizing systems. It has the ability for the device to learn from its own experience without human intervention. In the manufacturing industry, artificial intelligence can be used to analyze data from airborne sensors and perform preventive maintenance and process improvement to achieve autonomous artificial intelligence.

neural network 神经网络

deep learning 深度学习

self-organizing 自组织

AI can also play a role in optimizing the layout of workshops. Factory production should respond quickly to customer needs. In order to adapt to② the launch of new products, small batch production, process changes and other new situations, workshop equipment needs to be frequently reconfigured. The layout of the workshop takes into account③ factors ranging from the safety of the operator to the efficiency of the complete process. In order to make rational use of space and avoid conflicts, sensors are used for tracking and measurement to carry out

airborne sensor 机载遥感器

preventive maintenance 预防性维护

intelligent production, which requires artificial intelligence technology to ensure production efficiency and safety.

Artificial intelligence plays an important role in derivative design. Derivative design is a process of design exploration. Engineers input design objectives into derivative design software, along with product performance requirements and space requirements, as well as parameters such as materials, manufacturing methods, and cost constraints. Design software explores all possible combinations to quickly generate design alternatives. A valid solution is selected from each iteration through repeated testing, and the design is then distributed to multiple factories for subcontracting execution using compatible tools. In the automotive industry, for example, smaller, geographically dispersed partner plants can also produce more types of parts, thereby saving on④ subcontracting and shipping costs. Derivative design is becoming an important concept in the manufacturing industry.

In the era of Industry 4.0, artificial intelligence can be applied to innovative design, process improvement, equipment wears reduction, and energy consumption optimization in manufacturing. Machines are getting smarter and smarter. Automation is also becoming more integrated between equipment, finished products and supply chains, and other businesses. The idea is to free technicians from repetitive tasks that can be replaced by automation so that they can focus on innovation to create new ways of designing and manufacturing components.

Another key application area of AI in manufacturing is predictive maintenance. Engineers run pre-trained artificial intelligence models virtually, accumulating data and understanding of the production operation, and these models can use machine learning to discover causal patterns in the field and deal with problems in a timely manner to prevent actual production problems.

With complementary technologies such as virtual reality (as shown in Fig. 3-3) and augmented reality (as shown in Fig. 3-4), AI can provide solutions that shorten design time and optimize production and assembly line processes. Equipped with virtual and augmented reality technology systems, workers on the production line can see the production and assembly process, providing visual guidance to improve speed and accuracy.

derivative design
衍生式设计

energy consumption
能源消耗

virtual reality
（计算机创造的）
虚拟现实

augmented reality
增强现实

assembly line
装配线

Fig. 3-3　The Virtual Reality in Workshop

Fig. 3-4　The Augmented Reality in Workshop

 Tips

1. VR: Virtual Reality 虚拟现实

虚拟现实,又称虚拟环境、灵境或人工环境。虚拟现实技术是指利用计算机生成一种可对参与者直接施加视觉、听觉和触觉感受,并允许其交互观察和操作的虚拟世界的技术。虚拟现实系统的基本特征为:沉浸(immersion)、交互(interaction)和想象(imagination)。虚拟现实强调人在虚拟现实系统中的主导作用,使信息处理系统适合人的需要,并与人的感官感觉相一致。

2. AR: Augmented Reality 增强现实

增强现实技术是一种将真实世界信息和虚拟世界信息"无缝"集成的新技术。该技术是指把原本在现实世界的一定时间、空间范围内很难体验到的实体信息(视觉信息、声音、味道、触觉等)通过计算机技术等模拟仿真后叠加,将虚拟的信息应用到真实世界,使其被人类感官所感知,从而获得超越现实的感官体验。真实的环境和虚拟的物体实时地叠加到了同一个画面或空间并同时存在。

New Words and Phrases

detect [dɪˈtekt] v. 查明, 检测, 发现, 察觉
wear [weə(r)] v. 磨损, 穿(衣服)
fault [fɔːlt] n. 故障, 过失, 缺点, 缺陷
predict [prɪˈdɪkt] v. 预测, 预言
react [riˈækt] v. 反应, 回应
accurate [ˈækjərət] adj. 精确的, 准确的
troubleshooting [ˈtrʌblʃuːtɪŋ] n. 解决难题, 处理重大问题
model [ˈmɒdl] n. 样式, 设计, 模型
workpiece [ˈwɜːkpiːs] n. 工件
image [ˈɪmɪdʒ] n. 影像, 形象, 印象, 声誉
package [ˈpækɪdʒ] v. 将……包装好, 包装成
workflow [ˈwɜːkfləʊ] n. 工作流程
anomaly [əˈnɒməli] n. 异常, 异常事物, 反常现象
feedback [ˈfiːdbæk] n. 反馈, 反馈的意见(或信息)
parameter [pəˈræmɪtə(r)] n. 参数
trace [treɪs] v. 追踪, 查出, 追溯找到, 发现
load [ləʊd] n. 负载, 负荷, 装载量 v. 承载, 装入
intervention [ˌɪntəˈvenʃn] n. 干涉, 干预
optimize [ˈɒptɪmaɪz] v. 优化, 使最优化
layout [ˈleɪaʊt] n. 布局, 布置
workshop [ˈwɜːkʃɒp] n. 车间
launch [lɔːntʃ] n. (产品的)上市
constraint [kənˈstreɪnt] n. 限制, 限定, 约束
alternative [ɔːlˈtɜːnətɪv] adj. 可供替代的, 另类的, 非传统的 n. 可供选择的事物
valid [ˈvælɪd] adj. 有效的
solution [səˈluːʃn] n. 解决方案, 溶液
iteration [ˌɪtəˈreɪʃn] n. 迭代
subcontract [ˌsʌbˈkɒntrækt] v. 转包, 分包
compatible [kəmˈpætəbl] adj. 可共用的, 兼容的
disperse [dɪˈspɜːs] v. 分散, 散布, 疏散, 驱散
predictive maintenance 预见性维修
accumulate [əˈkjuːmjəleɪt] v. 积累, 积聚
complementary [ˌkɒmplɪˈmentri] adj. 互补的, 补充的, 相互补足的
machine tool 机床
process diagnosis 过程诊断

computer-aided design 计算机辅助设计
detection [dɪˈtekʃn] n. 检测，察觉，发现
sensor detection 传感器检测
nondestructive [ˌnʌndɪˈstrʌktɪv] adj. 无损的
nondestructive testing 无损检测
raw material 原材料，原料
finished part 成品零件
neural network 神经网络
deep learning 深度学习
self-organizing 自组织
airborne [ˈeəbɔːn] adj. 在空中的，飞行中的
airborne sensor 机载遥感器
preventive [prɪˈventɪv] adj. 预防性的
preventive maintenance 预防性维护
derivative [dɪˈrɪvətɪv] adj. 衍生的
derivative design 衍生式设计
consumption [kənˈsʌmpʃn] n. 消耗
energy consumption 能源消耗
virtual reality 虚拟现实
augmented [ɔːgˈmentɪd] adj. 增广的，增强的
augmented reality 增强现实
assembly [əˈsembli] n. 组装，装配
assembly line 装配线

Notes

1. give rise to 引起

例句：Artificial intelligence technology has given rise to "boxed factories" in manufacturing.

人工智能技术催生了制造业中的"盒装工厂"。

例句：The technological advances gave rise to the industrial revolution.

技术进步引发了工业革命。

2. adapt to 适应

例句：In order to adapt to the launch of new products, small batch production, process changes and other new situations, workshop equipment needs to be frequently reconfigured.

为了适应新产品上线、小批量生产、工艺流程变化等新情况，车间设备需要频繁地

重新配置。

例句：The world will be different, and we will have to be prepared to adapt to the change.

世界会变得不同，我们必须做好准备以适应其变化。

3. take into account 考虑到，把……计算在内

例句：The layout of the workshop takes into account factors ranging from the safety of the operator to the efficiency of the complete process.

车间布局要整体考虑从操作员的工作安全到完整工艺流程的效率等多种因素。

例句：Coursework is taken into account as well as exam results.

除考试结果外，课程作业也要计入成绩。

4. save on 节省，节约

例句：In the automotive industry, for example, smaller, geographically dispersed partner plants can also produce more types of parts, thereby saving on subcontracting and shipping costs.

例如，在汽车制造行业，规模较小、地理位置较分散的伙伴工厂也能生产出较多型号的零部件，从而节约了分包和运输成本。

例句：I save on fares by walking to work.

我步行上班，可以省车钱。

Exercises

Ⅰ. Write True or False beside the following statements about the text.

1. _____ Intelligent machine tools can automatically detect tool wear or faults, and even predict faults that will occur, react and solve them.

2. _____ Digital twin technology is an accurate digital representation of the processing.

3. _____ The idea behind the "boxed factory" process is that technicians load materials at one end and take finished parts out at the other.

4. _____ There is no need to reconfigure workshop equipment to adapt to new situations.

5. _____ It is hoped that technicians can be freed from repetitive tasks that can be replaced by automation to focus on innovation.

Ⅱ. Answer the following questions in English according to the text.

1. What can intelligent machine tools do?

2. What can users do?

3. What can artificial intelligence be used to do in the manufacturing industry?

4. What factors should be taken into account when laying out the workshop?

Ⅲ. Read the text again and fill in the blanks in the following sentences orally.

1. Artificial intelligence (AI) in manufacturing is the intelligence in which machines perform _____ tasks autonomously and _____ _____ events inside and around the machine.

2. The idea behind the "boxed factory" process is to load _____ _____ at one end and take _____ _____ out at the other, leaving _____ to maintain the system.

3. The realization of artificial intelligence comes largely from _____ _____, _____ _____, _____ _____, and other self-organizing systems.

4. Artificial intelligence plays an important role in _____ _____ which is a process of design exploration.

5. With complementary technologies such as _____ _____ and _____ _____, AI can provide solutions that shorten _____ _____ and optimize _____ and assembly line processes.

Ⅳ. Translation.

With complementary technologies such as virtual reality and augmented reality, AI can provide solutions that shorten design time and optimize production and assembly line processes. Equipped with virtual and augmented reality technology systems, workers on the production line can see the production and assembly process, providing visual guidance to improve speed and accuracy.

Lesson 4 Application of RFID Technology in Intelligent Manufacturing

Radio frequency identification (RFID) uses electromagnetic fields to automatically identify and track tags attached to objects. An RFID system consists of① a tiny radio transponder, a radio receiver, and a transmitter. When triggered by an electromagnetic interrogation pulse from a nearby RFID reader device, the tag transmits digital data, usually an identifying inventory number, back to the reader. This number can be used to track inventory goods.

Passive tags are powered by energy from the RFID reader's interrogating radio waves. Active tags are powered by a battery and thus can be read at a greater range from the RFID reader, up to hundreds of meters.

Margin Note

radio frequency identification
射频识别
electromagnetic field
电磁场
radio transponder
无线电应答器
radio receiver
无线电接收机

Unlike a barcode, the tag does not need to be within the line of sight of the reader, so it may be embedded in the tracked object. RFID is one method of automatic identification and data capture (AIDC).

RFID tags are used in many industries. For example, an RFID tag attached to[②] an automobile during production can be used to track its progress through the assembly line. RFID-tagged pharmaceuticals can be tracked through warehouses.

An RFID system uses tags, or labels attached to the objects to be identified. Two-way radio transmitter-receivers called interrogators or readers send a signal to the tag and read its response (as shown in Fig. 4-1).

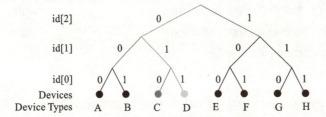

Fig. 4-1 An Example of a Binary Tree Method of Identifying an RFID Tag

RFID tags are made out of three pieces: a microchip (an integrated circuit that stores and processes information, modulates and demodulates radio frequency signals), an antenna for receiving and transmitting the signal, and a substrate. The tag information is stored in non-volatile memory. The RFID tag includes either fixed or programmable logic for processing the transmission and sensor data respectively.

RFID tags (as shown in Fig. 4-2) can be either passive, active, or battery-assisted passive. An active tag has an onboard battery and periodically transmits its ID signal. A battery-assisted passive tag has a small battery on board and is activated in the presence of[③] an RFID reader. A passive tag is cheaper and smaller because it has no battery. Instead, the tag uses the radio energy transmitted by the reader. However, to operate a passive tag, it must be illuminated with a power level that is roughly a thousand times stronger than an active tag for signal transmission. This can result in different levels of interference and radiation exposure.

RFID systems can be classified by the type of tag and reader. There are 3 types:

(1) A passive reader active tag (PRAT) system has a passive

---

radio transmitter
无线电发射机
electromagnetic interrogation pulse
电磁询问脉冲
reader device
阅读器
identifying inventory number
识别库存编号
inventory goods
库存
passive tag
无源标签
interrogating radio wave
询问无线电波
active tag
有源标签

two-way radio transmitter-receiver
双向无线电接发器

onboard battery
机载电池

Fig. 4-2　RFID Tag

reader which only receives radio signals from active tags (battery operated, transmit only).

(2) An active reader passive tag (ARPT) system has an active reader which transmits interrogator signals and also receives authentication replies from passive tags.

(3) An active reader active tag (ARAT) system uses active tags activated with an interrogator signal from the active reader.

Fixed readers are set up to create a specific interrogation zone that can be tightly controlled. This allows for a highly defined reading area when tags go in and out of the interrogation zone. Mobile readers may be handheld or mounted on carts or vehicles.

With the help of RFID technology, the whole life cycle management of products can be realized, and manufacturing efficiency and quality can be effectively improved. For example, a digital workshop based on RFID technology mainly includes tool accessory management, intelligent maintenance of equipment, and workshop mixed-flow manufacturing. RFID technology can be used to realize information interaction between tool accessories and host, visual tracking management of tool accessories, quantitative monitoring, and prediction of tool accessories' life. And the effective integration of RFID technology and sensor technology can quickly obtain the status information of products in processing, assembly, service, and other stages in real-time. At the same time, through network transmission, it provides powerful data support for background service support, remote instructions, and improvement in personalized design for users. Through the application of RFID technology, the production cycle and delivery time of enterprises are greatly shortened.

Tips

1. Barcode 条形码

条形码是将宽度不等的多个黑条和白条,按照一定的编码规则排列,以表达一组信息的图形标识符。常见的条形码是由反射率相差很大的黑条(简称条)和白条(简称空)排成的平行线图案。条形码可以标出物品的生产国、制造厂家、商品名称、生产日期、图书分类号、邮件起止地点、类别、日期等许多信息,因而在商品流通、图书管理、邮政管理、银行系统等许多领域得到广泛的应用。

New Words and Phrases

electromagnetic [ɪˌlektrəʊmæɡˈnetɪk] *adj.* 电磁的
electromagnetic field 电磁场
object [ˈɒbdʒɪkt] *n.* 物体 *v.* 不同意,不赞成,反对
tiny [ˈtaɪni] *adj.* 极小的,微小的,微量的
radio [ˈreɪdiəʊ] *n.* 无线电传送,收音机,广播电台
receiver [rɪˈsiːvə(r)] *n.* 无线电接收机
transmitter [trænzˈmɪtə(r)] *n.* (尤指无线电或电视信号的)发射机
trigger [ˈtrɪɡə(r)] *v.* 触发 *n.* 触发器
pulse [pʌls] *n.* 脉冲
tag [tæɡ] *n.* 标签 *vt.* 给……加上标签,把……称作
binary [ˈbaɪnəri] *adj.* 二进制的(用 0 和 1 记数),二元的,由两部分组成的
power [ˈpaʊə(r)] *v.* 驱动
battery [ˈbætri] *n.* 电池
barcode [ˈbɑːkəʊd] *n.* 条形码
sight [saɪt] *n.* 视力范围,视力,看见,视野
embed [ɪmˈbed] *v.* 嵌入
capture [ˈkæptʃə(r)] *v.* 捕获
pharmaceutical [ˌfɑːməˈsuːtɪkl] *n.* 药物 *adj.* 制药的
label [ˈleɪbl] *n.* 标签 *v.* 贴标签于,用标签标明
attach [əˈtætʃ] *v.* 贴上,把……固定,附上
periodically [ˌpɪəriˈɒdɪkəli] *adv.* 定期地,周期性地
exposure [ɪkˈspəʊʒə(r)] *n.* 暴露

radiation [ˌreɪdi'eɪʃn] n. 辐射
classify ['klæsɪfaɪ] v. 分类,划分,将……分类
authentication [ɔːˌθentɪ'keɪʃn] n. 身份验证,认证,鉴定
activate ['æktɪveɪt] v. 激活
tightly ['taɪtli] adv. 紧紧地,牢固地,紧密地
cart [kɑːt] n. 手推车,手拉车
vehicle ['viːəkl] n. 交通工具,车辆
accessory [ək'sesəri] n. 附件,配件,附属物
interaction [ˌɪntər'ækʃn] n. 相互影响,相互作用
visual ['vɪʒuəl] adj. 视觉的,视力的,可见的
transmission [trænz'mɪʃn] n. 播送,发射,发送,传输
remote [rɪ'məʊt] adj. 远程的,远程连接的,偏远的,偏僻的
frequency ['friːkwənsi] n. 频繁,频率
radio frequency identification 射频识别
transponder [træn'spɒndə(r)] n. 应答器,转发器
radio transponder 无线电应答器
radio receiver 无线电接收机
radio transmitter 无线电发射机
interrogation [ɪnˌterə'geɪʃn] n. 询问
electromagnetic interrogation pulse 电磁询问脉冲
interrogating radio wave 询问无线电波
reader device 阅读器
inventory ['ɪnvəntri] n. 存货,库存
identifying inventory number 识别库存编号
passive ['pæsɪv] adj. 无源的,被动的
passive tag 无源标签
active ['æktɪv] adj. 有源的,活跃的
active tag 有源标签
inventory goods 库存
onboard battery 机载电池
automatic identification and data capture 自动识别和数据捕获

 Notes

1. consist of 包括,由……组成(构成)

例句:An RFID system <u>consists of</u> a tiny radio transponder, a radio receiver, and a transmitter.

RFID 系统由一个微型无线电应答器、一个无线电接收机和发射机组成。

例句：Sunlight consists of radiation of different wavelengths.

阳光由几种不同波长的射线组成。

2. attach...to... 把……固定，贴上

例句：An RFID tag attached to an automobile during production can be used to track its progress through the assembly line.

在汽车生产过程中，贴在汽车上的 RFID 标签可以用来跟踪其在装配线上的进度。

例句：We attach labels to things before we file them away.

物品存档前，我们先贴上标签。

3. in the presence of 存在……的情况下，有……的存在

例句：A battery-assisted passive tag has a small battery on board and is activated in the presence of an RFID reader.

电池辅助无源标签上有一个小电池，当出现 RFID 阅读器时其会被激活。

例句：Litmus paper turns red in the presence of acid.

石蕊试纸遇到酸就变红。

Exercises

Ⅰ. Write True or False beside the following statements about the text.

1. _____ An RFID system consists of a tiny radio transponder, a radio receiver, and a transmitter.

2. _____ Passive tags are powered by a battery.

3. _____ The tag needs to be within the line of sight of the reader like a barcode.

4. _____ An active reader passive tag (ARPT) system uses active tags activated with an interrogator signal from the active reader.

5. _____ The production cycle and delivery time of enterprises are greatly shortened due to the application of RFID technology.

Ⅱ. Answer the following questions in English according to the text.

1. What are the differences between passive tags and active tags?

2. What can an RFID tag attached to an automobile during production be used to do?

3. What is an RFID tag made out of?

4. What are the benefits of RFID technology?

Ⅲ. Read the text again and fill in the blanks in the following sentences orally.

1. Radio frequency identification (RFID) uses _____ _____ to automatically _____ and _____ tags attached to objects.

2. RFID tags are made out of three pieces: _____ _____ _____, _____

_____ for receiving and transmitting the signal, and _____ _____.

3. A _____ passive tag has a small battery _____ _____ and is activated _____ _____ _____ _____ an RFID reader.

4. Fixed readers are _____ _____ to create a specific _____ _____ that can be tightly controlled.

5. With the help of _____ _____, the whole life cycle management of products can be realized, and manufacturing _____ and _____ can be effectively improved.

IV. Translation.

RFID tags are made out of three pieces: a micro chip (an integrated circuit that stores and processes information, modulates and demodulates radio frequency signals), an antenna for receiving and transmitting the signal and, a substrate. The tag information is stored in non-volatile memory. The RFID tag includes either fixed or programmable logic for processing the transmission and sensor data respectively.

Lesson 5   Intelligent Robot

Robots are particularly suitable for those occasions that put humans in danger, such as space exploration and search, rescue operations in disaster areas, and where, economically speaking, complex and repetitive work with high precision is required, such as production lines.

To Oxford English Dictionary: Robot is "... a machine designed to function in place of[①] a living agent, especially one which carries out[②] a variety of tasks automatically or with a minimum of external impulse".

Robotics is the discipline that involves:

(1) the design, manufacture, control, and programming of robots.

(2) the use of robots to solve problems.

(3) the study of the control processes, sensors, and algorithms used in humans, animals, and machines.

(4) the application of these control processes and algorithms to the design of robots.

There are some differences between a robot and a computer. Input symbols are static and well-behaved for a computer, whereas robot sensory signals are noisy and unreliable. Computer operations give consistent results, whereas a robot's action can have different responses. A computer's environment is fixed, whereas objects may move about independently in a robot's environment. A computer only

Margin Note

external impulse
外界冲击强度

input symbol
输入符号
robot sensory signal
机器人感知信号

receives the inputs intended for it, whereas a robot may have interference from various sources. In a computing environment, perfect performance is assumed; for a robot, its operating environment is unreliable, dynamic, and incomplete (as shown in Fig. 5-1).

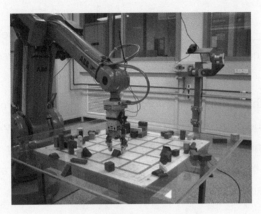

Fig. 5-1　Example of a Robot

An intelligent robot is a machine able to extract information from its environment and use knowledge about its world to move safely in a meaningful and purposive manner. Autonomy is the ability of a system to get on with its task without outside help. The development of robots can be defined as the process from non-autonomous to autonomous:

(1) Teleoperation — a robot is completely under the control of a person or people, such as robots in "Robot Wars".

(2) Telepresence — the use of communication and computing technology to allow a person to control robots and monitor the results from any distance.

(3) Semi-autonomy — supervisory control where human participates in partial control or control exchange.

(4) Fully autonomous — able to complete a task with no human input at all.

The dream of researchers in robotics is to create the ultimate human-like robot (as shown in Fig. 5-2), irrespective of[3] whether they admit it or not, and whatever moral implications and complications this might bring. But this ultimate robot will not be here tomorrow! However, all of its basic building blocks are there already: mechanical skeletons, motors and sensors, computer brains, reasoning, and control algorithms. Researchers all over the world can start constructing the "DNA" of the humanoid robot. All we need is time and a lot of effort to make it grow and reach its full potential.

control algorithm
控制算法

Fig. 5-2　Human-like Robot

 Tips

1. DNA 脱氧核糖核酸

脱氧核糖核酸是分子结构复杂的有机化合物,作为染色体的一个成分而存在于细胞核内,功能为储存遗传信息。1953 年,美国的沃森(Watson)、英国的克里克(Crick)与韦尔金斯(Wilkins)描述了 DNA 的结构:由一对多脱氧核苷酸链围绕一个共同的中心轴盘绕构成。脱氧核糖-磷酸链在螺旋形结构的外面,碱基朝向里面。两条多脱氧核苷酸链通过碱基之间的氢键相连,形成相当稳定的组合。

 New Words and Phrases

intelligent [inˈtelidʒənt] *adj*. 聪明的,有智力的,智能的
robot [ˈrəubɔt] *n*. 机器人,自动操作装置,机器般的人
exploration [ˌeksplɔːˈreiʃən] *n*. 探索,研究
space exploration 太空探索,外层空间探索
precision [prɪˈsɪʒn] *n*. 精确,准确,细致
production [prəˈdʌkʃən] *n*. 生产,加工,产量
production line 生产线
rescue [ˈreskjuː] *v*. 解救,援(营、挽)救,救出
consistent [kənˈsɪstənt] *adj*. 一致的,坚持的,坚固的
extract [ɪkˈstrækt] *v*. 提取,提炼
teleoperation [ˌteliˌɒpəˈreɪʃn] *n*. 远程操作

telepresence ['teliprezns] *n*. 遥现,远程在位(利用计算机模拟过程)
semi-autonomy 半自动,半自主
communication [kəmju:nɪ'keɪʃ(ə)n] *n*. 通讯,通信,交通,联络
skeleton ['skelitən] *n*. 骨骼,骨架,提纲
mechanical skeleton 机械框架,机械骨架
potential [pə'tenʃ(ə)l] *n*. 潜力,潜能,可能性 *adj*. 潜在的,可能的
impulse ['ɪmpʌls] *n*. 脉冲,冲动
external impulse 外界冲击强度
symbol ['sɪmbl] *n*. 符号,象征,标志,代表性的人(物)
input symbol 输入符号
robot sensory signal 机器人感知信号
algorithm ['ælgərɪðm] *n*. 算法,运算法则
control algorithm 控制算法

Notes

1. in place of 代替

例句:Robot is "... a machine designed to function in place of a living agent, especially one which carries out a variety of tasks automatically or with a minimum of external impulse".

机器人是……用于替代人类与动物作用的机器,尤其是能自动执行各种任务或以少量的外部指令完成各种任务的机器。

例句:In place of our advertised program, we will show a film.
我们将放映一部电影来代替我们的广告节目。

2. carry out 实现,完成,实行

例句:Robot is "... a machine designed to function in place of① a living agent, especially one which carries out② a variety of tasks automatically or with a minimum of external impulse".

机器人是……用于替代人类与动物作用的机器,尤其是能自动执行各种任务或以少量的外部指令完成各种任务的机器。

例句:I have carried out my work.
我已经完成了我的工作。

3. irrespective of 不顾……的,不考虑……的,不论……的

例句:The dream of researchers in robotics is to create the ultimate human-like robot, irrespective of whether they admit it or not, and whatever moral implications and complications this might bring.

不管他们是否承认,也不管这样做可能在道德上产生什么样的影响和纠纷,机器人

技术研究者们的梦想是最终创造出类人机器人。

例句：They send information every week, irrespective of whether it's useful or not.

他们每星期送出信息，不管其是否有用。

 Exercises

Ⅰ. Write True or False beside the following statements about the text.

1. _____ An intelligent robot is a machine able to get information from its environment and use knowledge about its world to move safely in a meaningful and purposive manner.

2. _____ In a computing environment, perfect performance is assumed; for a robot, its operating environment is reliable, dynamic, and complete.

3. _____ The development of robots can be defined as the process from autonomous to non-autonomous.

4. _____ Fully autonomous robots can be able to complete a task with no human input at all.

5. _____ Researchers all over the world can start constructing the "DNA" of the robot that looks and behaves like a human.

Ⅱ. Answer the following questions in English according to the text.

1. Where is a robot particularly suitable for use? Where would it be more economically appropriate to use robots?

2. Please compare a robot with a computer.

3. What is the definition of semi-autonomy?

Ⅲ. Read the text again and fill in the blanks in the following sentences orally.

1. A robot is a machine designed to function in place of _____ _____ _____, especially one which _____ _____ a variety of tasks automatically or with a minimum of external impulse.

2. According to the first paragraph, robots are particularly suitable for _____ _____ _____ _____, _____ _____ _____ _____ _____, and _____ _____ _____.

3. Computer operations give consistent results, whereas a robot's action can have _____ _____.

4. Compared with a computer, a robot, its operating environment is _____, dynamic, and _____.

5. However, all of its basic building blocks are there already: mechanical skeletons, motors and _____, _____ _____, reasoning and _____ _____.

Ⅳ. Translation.

The dream of researchers in robotics is to create the ultimate human-like robot, irrespective of whether they admit it or not, and whatever moral implications and complications this might bring. All of its basic building blocks are there already: mechanical skeletons, motors and sensors, computer brains, reasoning, and control algorithms. Researchers all over the world can start constructing the "DNA" of the humanoid robot. All we need is time and a lot of effort to make it grow and reach its full potential.

Reading Material 1　　Artificial Intelligence

Computer scientists have tried to develop techniques that would allow computers to act more like humans. The techniques including decision-making systems, robotic devices (as shown in Fig. R1-1 and Fig. R1-2), and various approaches to computer speech are usually called artificial intelligence (AI). A computer program is a set of instructions that enables a computer to process information and solve problems. And AI programs can take shortcuts, make choices, search for and try out different solutions, and change their methods of operation.

Margin Note

Fig. R1-1　Robotic Devices 1

Recently, chess-playing computer programs have been developed to defeat most human opponents including chess masters. Of course, there's more to artificial intelligence than the ability to play games. Computer scientists are working on dozens of different practical uses for AI programs including operating robots, solving math and science problems, understanding speech, and analyzing images.

Perhaps the biggest use of AI programs is expert advisors for troubleshooting (locating problems and making repairs) complex systems ranging from diesel engines to nuclear submarines and the

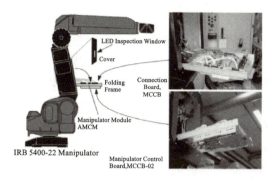

Fig. R1-2　Robotic Devices 2

human body. In other words, these AI programs search for trouble, detect and classify problem areas, and give advice. The main methods of AI systems that have been proposed for integration are message routing, or communication protocols that the software components used to communicate with each other, often through a middleware blackboard system (as shown in Fig. R1-3 and Fig. R1-4).

message routing
消息路由

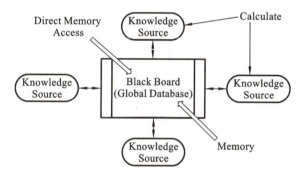

Fig. R1-3　A Middleware Blackboard System 1

Fig. R1-4　A Middleware Blackboard System 2

　　The traditional goals of AI research include reasoning, knowledge representation, planning, learning, natural language processing, perception, and the ability to move and manipulate objects. General intelligence (the ability to solve an arbitrary problem) is among the field's long-term goals. To solve these problems, AI researchers have adapted and integrated a wide range of problem-solving techniques including search and mathematical optimization, formal logic, artificial neural networks, and methods based on statistics, probability, and economics. AI also draws upon computer science, psychology, linguistics, philosophy, and many other fields.

  The applications of modern artificial intelligence techniques are pervasive and are too numerous to list here. In the 2010s, AI applications were at the heart of the most commercially successful areas of computing and had become a ubiquitous feature of daily life. AI is used in search engines, targeting online advertisements, recommendation systems, driving internet traffic, targeted advertising, virtual assistants, autonomous vehicles (drones and self-driving cars), automatic language translation, facial recognition, image labeling, and spam filtering.

  There are also thousands of successful AI applications used to solve problems for specific industries or institutions. A few examples are energy storage, deep fakes, medical diagnosis, military logistics, and supply chain management.

  Game playing has been a test of AI's strength since the 1950s. Deep Blue became the first computer chess-playing system to beat a reigning world chess champion, Garry Kasparov. AlphaGo won 4 out of 5 games of Go in a match with Go champion Lee Sedol, becoming the first computer Go-playing system to beat a professional Go player without handicaps.

  By 2020, natural language processing systems such as the enormous GPT-3 (then by far the largest artificial neural network) were matching human performance on pre-existing benchmarks.

  Artificial intelligence and the internet of things, big data technology, robotics, 3D printing, and fifth-generation (5G) mobile services are among the most prolific emerging technologies today.

  Major AI application fields include telecommunications, transportation, life and medical sciences, personal devices, computing, and human-computer interaction. Other sectors include banking, entertainment, security, industrial and manufacturing, agriculture, social networks, smart cities, and the internet of things.

> mathematical optimization 数学优化
> formal logic 形式逻辑
> artificial neural network 人工神经网络

New Words and Phrases

shortcut [ˈʃɔːtkʌt] n. 捷径,快捷方式(图标)
protocol [ˈprəʊtəkɒl] n. (数据传递的)协议,规程,规约
middleware [ˈmɪdlweə(r)] n. 中间件,中介软件(允许不同程序协同工作)
manipulate [məˈnɪpjuleɪt] v. (熟练地)操作,使用
statistics [stəˈtɪstɪks] n. 统计数字(或资料),统计学
probability [ˌprɒbəˈbɪləti] n. 概率,可能性
economics [ˌiːkəˈnɒmɪks] n. 经济学,经济情况
draw upon 利用
psychology [saɪˈkɒlədʒi] n. 心理学
linguistics [lɪŋˈɡwɪstɪks] n. 语言学
philosophy [fəˈlɒsəfi] n. 哲学
pervasive [pəˈveɪsɪv] adj. 遍布的,充斥各处的,弥漫的
ubiquitous [juːˈbɪkwɪtəs] adj. 似乎无所不在的,十分普遍的
spam filtering 垃圾邮件过滤
benchmark [ˈbentʃmɑːk] n. 基准
prolific [prəˈlɪfɪk] adj. 多产的,创作丰富的
message routing 消息路由
mathematical optimization 数学优化
formal logic 形式逻辑
artificial neural network 人工神经网络

Reading Material 2   Expert System

In artificial intelligence, an expert system is a computer system emulating the decision-making ability of human experts. Expert systems are designed to solve complex problems by reasoning through bodies of knowledge, represented mainly as if-then rules rather than through conventional procedural code. Expert systems were among the first truly successful forms of artificial intelligence (AI) software. An expert system is divided into two subsystems: the inference engine and the knowledge base. The system also includes the following components: an explanation facility, a knowledge acquisition facility, and a user interface. The knowledge base represents facts and rules. The inference

Margin Note

if-then rule
如果-那么规则

engine applies the rules to the known facts to deduce new facts. Inference engines can also include explanation and debugging abilities.

As shown in Fig. R2-1, the system is constituted of the application executive, user interface, diagnostic reasoner(s), test controller(s), and test and maintenance information management center.

application executive
应用执行器

user interface
用户界面

diagnostic reasoner(s)
诊断推理器

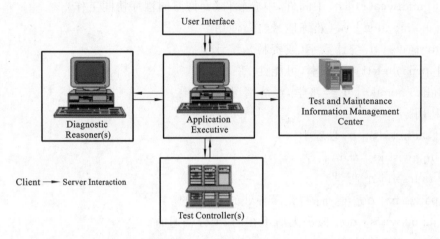

Fig. R2-1　An Expert System

As expert systems evolved, many new techniques are incorporated into various types of inference engines. Some of the most important of these are:

(1) Truth maintenance. These systems record the dependencies in a knowledge base so that when facts are altered, dependent knowledge can be altered accordingly. For example, if the system learns that Socrates is no longer known to be a man it will revoke the assertion that Socrates is mortal.

(2) Hypothetical reasoning. In this, the knowledge base can be divided up into many possible views, also known as worlds. This allows the inference engine to explore multiple possibilities in parallel. For example, the system may want to explore the consequences of both assertions, what will be true if Socrates is a man, and what will be true if he is not?

(3) Uncertainty systems. One of the first extensions of simply using rules to represent knowledge is also to associate a probability with each rule. So, not to assert that Socrates is mortal, but to assert Socrates may be mortal with some probability value. Simple probabilities are extended in some systems with sophisticated mechanisms for uncertain reasoning, such as fuzzy logic, and a combination of probabilities.

(4) Ontology classification. With the addition of object classes to the knowledge base, a new type of reasoning is possible. Along with reasoning simply about object values, the system could also reason about object structures. In this simple example, man can represent an object class and R1 can be redefined as a rule that defines the class of all men. These types of special-purpose inference engines are termed classifiers. Although they are not highly used in expert systems, classifiers are very powerful for unstructured volatile domains and are a key technology for the internet and the emerging semantic web.

semantic web
语义网

New Words and Phrases

emulate [ˈemjuleɪt] v. 仿真,模仿,同……竞争
conventional [kənˈvenʃənl] adj. 传统的,习惯的
procedural [prəˈsiːdʒərəl] adj. 程序上的,程序性的
inference [ˈɪnfərəns] n. 推断,推理,推论
acquisition [ˌækwɪˈzɪʃn] n.(知识、技能等的)获得,得到
dependency [dɪˈpendənsi] n. 依靠,依赖
alter [ˈɔːltə(r)] v.(使)改变,更改,改动
revoke [rɪˈvəʊk] v. 取消,废除,使无效
assertion [əˈsɜːʃn] n. 主张,声称
hypothetical [ˌhaɪpəˈθetɪkl] adj. 假设的,假定的
consequence [ˈkɒnsɪkwəns] n. 结果,后果
mortal [ˈmɔːtl] adj. 不能永生的,终将死亡的
sophisticated [səˈfɪstɪkeɪtɪd] adj. 复杂巧妙的,先进的,精密的
mechanism [ˈmekənɪzəm] n. 机构,结构,构造,机械装置
ontology [ɒnˈtɒlədʒi] n. 本体,本体论
volatile [ˈvɒlətaɪl] adj. 易变的,动荡不定的,反复无常的
domain [dəˈmeɪn] n. 域,定义域
if-then rule 如果-那么规则
application executive 应用执行器
user interface 用户界面
diagnostic reasoner(s) 诊断推理器
semantic web 语义网

Reading Material 3　Cloud Manufacturing

Cloud manufacturing system is a service-oriented, knowledge-based smart manufacturing system with high efficiency and low energy consumption. In a cloud manufacturing system, state-of-the-art technologies such as informatized manufacturing technology, cloud computing, internet of things, semantic web, high-performance computing, and cloud manufacturing are integrated. By extending and shifting existing manufacturing and service systems, manufacturing resources and capabilities are virtualized and oriented toward service provision. In cloud manufacturing, pervasive and efficient sharing and coordination of resources and capabilities can be achieved by their unified and centralized intelligent management and operation. Cloud manufacturing provides the whole manufacturing lifecycle with secure, reliable, high-quality, and on-demand services at low prices through a networked system. The manufacturing lifecycle includes pre-manufacturing (argumentation, design, production, and sale), manufacturing (product usage, management, and maintenance), and post-manufacturing (dismantling, scrap, and recycling).

A cloud manufacturing system consists of manufacturing resources and capabilities, manufacturing cloud, and the whole manufacturing lifecycle applications. It also includes core support (knowledge), two processes (import and export), and three user types (resource providers, cloud operators, and resource users). Fig. R3-1 illustrates the operational principle of cloud manufacturing. Manufacturing resources and capabilities are encapsulated as cloud services. This process is called manufacturing resource "import". Depending on different manufacturing requirements, cloud services are combined to form a manufacturing cloud. The cloud provides the whole manufacturing lifecycle of applications with diverse services. This process is called "export". Knowledge plays a central role in supporting the entire operating process of cloud manufacturing. It is necessary for intelligent embedding and virtualized encapsulation during import; it assists

Margin Note

intelligent
embedding
智能嵌入

functions such as intelligent search of cloud services, and it facilitates smart cooperation of cloud services over the whole manufacturing lifecycle. In a cloud manufacturing system, knowledge-based integration across the whole lifecycle is possible.

virtualized
encapsulation
虚拟封装

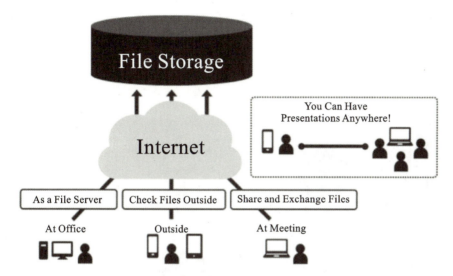

Fig. R3-1　Operational Principle of Cloud Manufacturing

Fig. R3-2 illustrates a cloud manufacturing application model. Users send specific requests to the cloud manufacturing platform. This platform is responsible for the management, operation, and maintenance of manufacturing clouds and service tasks such as import and export. It analyzes and divides service requests, and automatically searches the cloud for best-matched services. By a series of processes including scheduling, optimization, and combination, a solution is generated and then sent back to the client. A user does not need to communicate directly with every service node, nor find the specific locations and situations of service nodes. Through the cloud manufacturing platform, manufacturing resources and capabilities can be used in the same way as water, gas, electricity, etc.

Cloud manufacturing system architecture has five layers: physical layer, virtualized resource layer, service layer, application layer, and user layer.

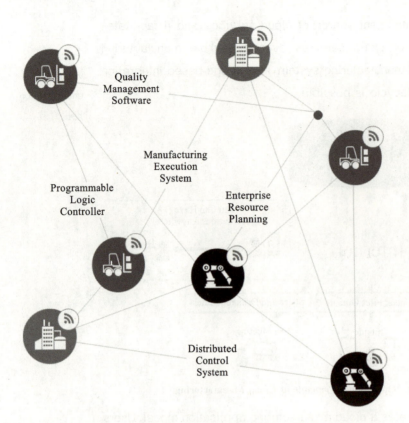

Fig. R3-2　Application Model of Cloud Manufacturing

 New Words and Phrases

state-of-the-art 使用最先进技术的,体现最高水平的
orient towards 朝向,面对,确定方向,使适应
provision [prə'vɪʒn] n. 提供,供给,给养,供应品
unify ['juːnɪfaɪ] v. 统一,使成一体,使一元化
dismantle [dɪs'mæntl] v. 拆开,拆卸
scrap [skræp] v. 废弃
recycle [ˌriː'saɪkl] v. 再次应用,重新使用,回收利用
encapsulate [ɪn'kæpsjuleɪt] v. 压缩,简述,概括
intelligent embedding 智能嵌入
architecture ['ɑːkɪtektʃə(r)] n. 体系结构,(总体、层次)结构
virtualized encapsulation 虚拟封装

Reading Material 4   Computer Vision

Computer vision means artificial sight by means of computers and other pertinent techniques. Compared with human sight, today's computer vision system is crude. However, it is a promising development direction with brilliant prospects for computer science and technology.

A typical real-time computer vision system includes the following components (as shown in Fig. R4-1).

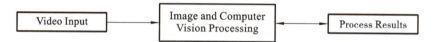

Fig. R4-1   Real-time Computer Vision System Components

1. Image Acquisition

A TV camera is usually used to take instantaneous images and transform them into electrical signals, which will be further translated into binary numbers for the computer to handle. The TV camera scans one line at a time. Each line is further divided into hundreds of pixels. The whole frame is divided into hundreds of lines (for example, 625 lines). The brightness of a pixel can be represented by a binary number with certain bits, for example, 8 bits. The value of the binary number varies from 0 to 255, a range great enough to accommodate all possible contrast levels of images taken from real scenes. These binary numbers are stored in RAM with a great capacity ready for further processing by the computer.

Margin Note

binary number
二进制数

RAM
随机存储器

2. Image Processing

Image processing is for improving the quality of the images obtained.

First, it is necessary to improve the signal-to-noise ratio. Here noise refers to any interference flaw or aberration that obscures the objects in the image. Second, it is possible to improve contrast, and enhance the sharpness of edges between images through various computational means.

3. Image Analysis

It is for outlining all possible objects that are included in the scene.

A computer program checks through the binary visual information in store for it and identifies specific features and characteristics of those objects. Edges or boundaries are identifiable because of the different brightness levels on either side of them. Using certain algorithms, the computer program can outline all possible boundaries of the objects in the scene. Image analysis also looks for textures and shadings between lines.

4. Image Comprehension

Image comprehension means understanding what is in a scene. Matching the prestored binary visual information with certain templates which represent specific objects in a binary form is a technique borrowed from artificial intelligence, commonly referred to as "template matching". One by one, the templates are checked against the binary information representing the scene. Once a match occurs, an object is identified. The template matching process continues until all possible objects in the scene have been identified, otherwise, it fails (as shown in Fig. R4-2).

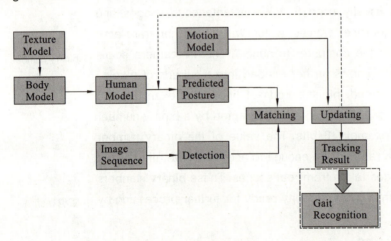

Fig. R4-2　Computer Vision

Computer vision has found its way into industries, doing jobs that used to be exclusive for human operators, for example, identifying specific objects or patterns by finding out distinctive details in shape, size, color, contrast, etc., inspecting products by checking for any flaws such as crack, smear. Jobs like these are boring and tiresome, and once the operator gets bored or tired, he/she tends to miss details or otherwise botch the manufacturing process.

One popular application of computer vision is machine vision.

Giving vision to machines can automate their manufacturing process with little or no human intervention. Machine vision greatly improves productivity and quality, and the cost and time of manufacture can be dramatically reduced.

The promising application of computer vision is in robots. If a robot is equipped with computer vision, it becomes an intelligent robot. A robot with a sight would allow it to adjust its operations to fit itself to varying conditions and environments.

Today, computer vision is extensively used in non-industrial fields as well, for example, for identifying fingerprints or facial features of a suspect, distinguishing counterfeit notes and forged paintings, analyzing medical images and photographs taken by satellites, etc.

New Words and Phrases

pertinent ['pɜːtɪnənt] *adj*. 有关的，恰当的，相宜的
crude [kruːd] *adj*. 粗略的，简略的，大概的
instantaneous [ˌɪnstən'teɪniəs] *adj*. 立即的，立刻的，瞬间的
pixel ['pɪksl] *n*. 像素
bit [bɪt] *n*. 比特
flaw [flɔː] *n*. 错误，缺点
aberration [ˌæbə'reɪʃn] *n*. 脱离常规，反常现象，异常行为
obscure [əb'skjʊə(r)] *v*. 使晦涩，使费解，使难懂
outline ['aʊtlaɪn] *v*. 显示……的轮廓，勾勒……的外形，概述，略述
boundary ['baʊndri] *n*. 边界，界限，分界线
template ['templeɪt] *n*. 模板
exclusive [ɪk'skluːsɪv] *adj*. 专用的，专有的，独有的，独占的
crack [kræk] *n*. 缝隙，狭缝，窄缝
smear [smɪə(r)] *n*. 污迹，污渍，污点
botch [bɒtʃ] *v*. 笨拙地弄糟
distinguish [dɪ'stɪŋɡwɪʃ] *v*. 区分，辨别，分清
binary number 二进制数

Unit II Industrial Robot Technology

 Introduction to the Unit 单元导言

> 机器人技术被视为变革制造业的主要技术,是各院校普遍开设的专业之一。本单元将重点讨论有关工业机器人技术应用方面的内容,主要是工业机器人、码垛机器人、自主移动机器人等的技术及应用。同时结合我国常用的几种机器人在生产一线的应用,学习操作手册中关于工业机器人的基本操作实训、机器人零件清单、机械手的安装、机械手的保养、机械手控制柜的维护、旋转计数器更新等操作层面的内容。通过本单元的学习,学生可以熟悉进口和出口机器人的英语产品说明书与操作手册,为毕业时顺利进入外向型企业就业以及走出国门深造或被派往海外工作打下基础。

 Lesson 6 Industrial Robot

An industrial robot is a robot system used for manufacturing. Industrial robots are automated, programmable, and capable of movement on three or more axes.

Typical applications of robots include welding, painting, assembly, disassembly, picking and placing of printed circuit boards, packaging and labeling, palletizing, product inspection, and testing. Industrial robots can also assist in[①] material handling. All these operations are completed with high durability, speed, and precision.

Articulated robots (as shown in Fig. 6-1) are the most common industrial robots. They look like human arms, which is why they are also called robotic arms or manipulator's arms. Their articulations with several degrees of freedom allow the articulated arms a wide range of[②] movements.

Cartesian robots (as shown in Fig. 6-2), also called rectilinear robots, gantry robots, or x-y-z robots, have three prismatic joints for the movement of the tool and three rotary joints for their orientation in space.

Margin Note

articulated robot
多关节型机器人
manipulator arm
机械手臂
degree of freedom
自由度
Cartesian robot
直角坐标型机器人
rectilinear robot
直线机器人

Fig. 6-1 An Articulated Robot   Fig. 6-2 A Cartesian Robot

The cylindrical coordinate robots are characterized by their rotary joint at the base and at least one prismatic joint connecting their links. They can move vertically and horizontally by sliding. The compact effector design allows the robot to reach a tight workspace without any loss of speed.

To be able to move and orient the effector organ in all directions, an x-y-z robot has 6 axes (or degrees of freedom) (as shown in Fig. 6-3), three degrees of freedom to move along the x, y, and z axes, and three degrees of freedom to rotate around the x, y, and z axes. In a 2-dimensional environment, three axes are sufficient, two for displacement and one for orientation. A seven degrees of freedom polishing manipulator has been developed (as shown in Fig. 6-4).

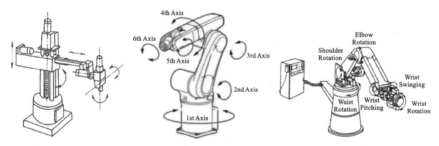

Fig. 6-3 Six DoF for a Robot and a Six-axis Robot for Spraying

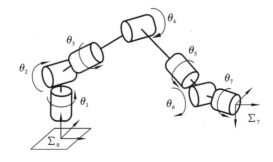

Fig. 6-4 The Seven Degrees of Freedom of a Bionic Human Arm

Delta robots are also referred to as parallel link robots. They consist of parallel links connected to a common base. Delta robots are particularly useful for direct control tasks and high maneuvering operations (such as quick pick-and-place tasks). Delta robots take advantage of four-bar or parallelogram linkage systems (as shown in Fig.6-5).

parallelogram linkage system
平行四边形连杆系统

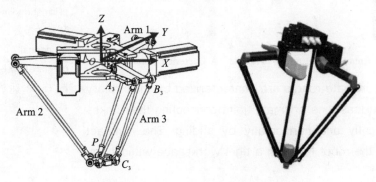

Fig. 6-5　Four-bar or Parallelogram Linkage Systems

Furthermore, industrial robots can have a serial or parallel architecture. A parallel manipulator (as shown in Fig. 6-6) is designed so that each chain is usually short and simple, and can thus be rigid against the unwanted movement, compared to[3] a serial manipulator. Errors in one chain's positioning are averaged in conjunction with[4] the others, rather than being cumulative. Each actuator must still move within its own degree of freedom, as for a serial robot; however, in the parallel robot, the off-axis flexibility of a joint is also constrained by the effect of the other chains. It is this closed-loop stiffness that makes the overall parallel manipulator stiff relative to its components, unlike the serial chain that becomes progressively less rigid with more components.

parallel manipulator
并联机械手

serial manipulator
串联机械手

closed-loop stiffness
闭环刚性

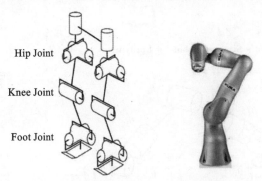

Fig. 6-6　A Leg and Arm Bionic Flexible Parallel Robot

Robots exhibit varying degrees of autonomy. Some robots are programmed to faithfully carry out specific actions over and over again (repetitive actions) without variation and with a high degree of accuracy. These actions are determined by programmed routines that

programmed routine
例行程序

specify the direction, acceleration, velocity, deceleration, and distance of a series of coordinated motions.

Other robots are much more flexible as to the orientation of the object on which they are operating or even the task that has to be performed on the object itself, which the robot may even need to identify. For example, for more precise guidance, robots often contain machine vision sub-systems acting as their visual sensors, linked to powerful computers or controllers. Artificial intelligence is becoming an increasingly important factor in the modern industrial robot.

visual sensor 视觉传感器

 Tips

1. Articulated Robot 多关节型机器人

多关节型机器人与人的手臂类似，其特点是能像人手那样灵活动作。如果在多关节型机器人手部和腕部装上触觉和力的传感器，它就能做更多、更复杂的工作。多关节型机器人是当今工业领域中常见的工业机器人之一，适合用于诸多工业领域的机械自动化作业。

2. Cartesian Robot 直角坐标型机器人

直角坐标型机器人是指在工业应用中，能够实现自动控制的、可重复编程的、运动自由度仅包含三维空间正交平移的自动化设备。其由直线运动轴、运动轴的驱动系统、控制系统、终端设备等组成。各个运动轴通常对应直角坐标系中的 X 轴、Y 轴和 Z 轴。在一些应用中 Z 轴上带一个旋转轴，或带一个摆动轴和一个旋转轴。直角坐标型机器人可在多领域中被应用，具有超大行程、强组合能力等优点。

3. Cylindrical Coordinate Robot 圆柱坐标型机器人

圆柱坐标型机器人以 θ、z 和 r 为参数构成坐标系。手腕参考点的位置可表示为 $P=f(\theta,z,r)$。其中，r 是手臂的径向长度，θ 是手臂绕水平轴的角位移，z 是在垂直轴上的高度。如果 r 不变，则操作臂的运动将形成一个圆柱表面，空间定位比较直观。

4. Delta Robot 三角机器人

三角机器人是高精度、高效率、长寿命的并联机器人，主要用于食品包装行业的抓取、包装、码垛和机床上下料，具有重量轻、体积小、运动速度快、定位精确等特点。三角机器人通过网络即可完成远程监控维护，操作简单，可完美地实现生产数字化、自动化、网络化以及智能化。

New Words and Phrases

programmable [ˈprəʊɡræməbl] adj. 可以编程的，计算机程序控制的
axis [ˈæksɪs] n. 轴（旋转物体假想的中心线）
weld [weld] v. 焊接，熔接，锻接
disassembly [ˌdɪsəˈsembli] n. 拆卸，分解
pick and place 拾取与放置，贴装，贴片，取放
palletize [ˈpælətaɪz] v. 码垛堆积
durability [ˌdjʊərəˈbɪlɪti] n. 耐久性
rotary [ˈrəʊtəri] adj. 旋转的，绕轴转动的，转动的
vertically [ˈvɜːtɪkəli] adv. 垂直地，直立地
horizontally [ˌhɒrɪˈzɒntəli] adv. 水平地，横地
slide [slaɪd] v. （使）滑行，滑动
spray [spreɪ] v. 喷，喷洒，向……喷洒
bionic [baɪˈɒnɪk] adj. （因体内有电子装置）能力超人的
compact [ˈkɒmpækt] adj. 紧凑的，紧密的，体积小的
tight [taɪt] adj. 装紧的，密集的，挤满的
displacement [dɪsˈpleɪsmənt] n. 移位，取代
polish [ˈpɒlɪʃ] v. 擦光，磨光
maneuver [məˈnuːvə] v. （熟练地）移动，调动，转动，操纵
average [ˈævərɪdʒ] v. 平均为 adj. 平均的 n. 平均数
cumulative [ˈkjuːmjələtɪv] adj. 聚积的，积累的，渐增的
actuator [ˈæktjʊeɪtə] n. 执行机构（元件）
acceleration [əkˌseləˈreɪʃn] n. 加速，加快
deceleration [ˌdiːseləˈreɪʃn] n. 减速，降速
sub-system 子系统
articulated [ɑːˈtɪkjuleɪtɪd] adj. 有关节的，铰接的
articulated robot 多关节型机器人
manipulator arm 机械手臂
Cartesian robot 直角坐标型机器人
rectilinear [ˌrektɪˈlɪnɪə(r)] adj. 直线运动的
rectilinear robot 直线机器人
gantry [ˈɡæntri] n. 桁架，桶架
gantry robot 桁架机器人
x-y-z robot 三坐标机器人
prismatic [prɪzˈmætɪk] adj. 棱柱的，棱镜的
prismatic joint 移动关节

cylindrical [sə'lɪndrɪkl] adj. 圆柱体的
cylindrical coordinate robot 圆柱坐标型机器人
effector [ɪ'fektə(r)] n. 效应器
effector organ 操纵机构
degree of freedom 自由度
delta robot 三角机器人
parallelogram [ˌpærə'leləɡræm] n. 平行四边形
parallelogram linkage system 平行四边形连杆系统
parallel manipulator 并联机械手
serial ['sɪəriəl] adj. 串联的，串行的
serial manipulator 串联机械手
stiffness [stɪfnəs] n. 刚度，硬度

Notes

1. assist in 帮助，协助，援助

例句：Industrial robots can also assist in material handling.
工业机器人还可以协助物料搬运。

例句：We are looking for people who would be willing to assist in the group's work.
我们正寻找愿意协助该团体工作的人。

2. a range of 范围，射程

例句：Their articulations with several degrees of freedom allow the articulated arms a wide range of movements.
它们的关节具有多个自由度，使得关节臂能够进行大范围的运动。

例句：The hotel offers a wide range of facilities.
这家酒店提供各种各样的设施。

3. compared to 与……相比

例句：A parallel manipulator is designed so that each chain is usually short and simple, and can thus be rigid against the unwanted movement, compared to a serial manipulator.
与串联机械手相比，并联机械手的设计使每个链条通常短而简单，因此可以刚性地防止不必要的移动。

例句：I've had some difficulties, but they were nothing compared to yours.
我遇到了一些困难，但与你的困难比起来就算不上什么了。

4. in conjunction with 与……一起

例句：Errors in one chain's positioning are averaged in conjunction with the others, rather than being cumulative.

在结合处,一个连杆的定位误差与其他连杆的定位误差是均分的,而不是累积的。

例句:The system is designed to be used in conjunction with a word-processing program.

本系统是为与文字处理软件配合使用而设计的。

 Exercises

Ⅰ. Write True or False beside the following statements about the text.

1. _____ Industrial robots are automated, programmable, and capable of movement on no more than three axes.

2. _____ Rectilinear robots have three rotary joints for the movement of the tool and three prismatic joints for their orientation in space.

3. _____ The cylindrical coordinate robots can reach tight workspaces without any loss of speed.

4. _____ In a 2-dimensional environment, three axes DoF are sufficient, two for displacement and one for orientations.

5. _____ Robots can be programmed to carry out specific repetitive actions.

Ⅱ. Answer the following questions in English according to the text.

1. What can industrial robots do?

2. What is an articulated robot?

3. How many axes are needed to enable a robot to move in all directions?

Ⅲ. Read the text again and fill in the blanks in the following sentences orally.

1. Articulated robots are the most common industrial robots which _____ _____ human arms.

2. The cylindrical coordinate robots are _____ by their rotary _____ at the base and at least one prismatic joint _____ their links.

3. An x-y-z robot has 6 axes, so it can _____ and _____ the effector organ in all _____.

4. Compared to a _____ manipulator, a _____ manipulator is designed so that each chain is usually _____ and _____, and can thus be rigid against the unwanted _____.

5. Actions carried out by robots are _____ by programmed _____ that _____ the direction, _____, velocity, deceleration, and distance of a series of _____ motions.

Ⅳ. Translation.

To be able to move and orient the effector organ in all directions, an x-y-z robot has 6 axes, three degrees of freedom to move along the x, y, and z axes, and three degrees of freedom to rotate around the x,y,and z axes. In a 2-dimensional environment, three axes are sufficient, two for displacement and one for orientation.

Lesson 7　The Maintenance of the Manipulator

Table 7-1 lists manipulator maintenance. This manual is used for "quick reference" of the ABB robot manual. Please read the ABB robot manual in detail.

Margin Note

manipulator
机械手

Table 7-1　Manipulator Maintenance

No.	Activity and Reference	Interval
1	General Maintenance (Part 1)	Daily
2	Lubrication of Balance Drive Bearings (Part 2)	3 000 hours
3	Lubrication of Balance Spring Rods (Part 3)	3 000 hours
4	Lubrication of Hollow Wrists (Part 4)	3 000 hours
5	Check Oil Level in Gear Boxes for axis 1-6 (Part 5)	5 years

balance drive bearing
平衡传动轴承

balance spring rod
平衡弹簧杆

1. Part 1　General Maintenance

1) Clean the Manipulator

The manipulator (as shown in Fig. 7-1) base and arm assembly should be cleaned regularly. Solvents can be used but with care[①]. Strong solvents such as acetone should be avoided.

arm assembly
手臂组件

strong solvent
强溶剂

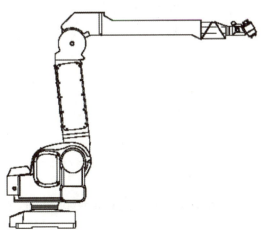

Fig. 7-1　The Manipulator

2) Clean/Check Paint Components

Clean the external contaminated paint parts with a wet cloth after each shift and during the break. After cleaning, carefully dry the components with pressurized air. Pressurized air and solvents can be used but with care. Strong solvents such as acetone should be avoided.

3) Clean Hollow Wrist

The wrist must be cleaned as often as required. And take care to avoid the accumulation of dirt and dust particles. Use lint-free cloth for cleaning. After the wrist is cleaned, a small amount of vaseline should be added to[2] the surface of the wrist. This will simplify the next cleaning.

hollow wrist
空心腕关节
lint-free cloth
无绒布
vaseline
凡士林

4) Check Regularly

For oil leaks, if serious oil leakage is found, call for[3] service personnel. If the gear rotates excessively, call for assistance. Make sure that the cables between the control cabinet, purging unit, and manipulator are not damaged.

service personnel
维修人员
purging unit
清洁装置

2. Part 2 Lubrication of Balance Drive Bearings

Caution: this maintenance is very important. It must be done every 3 000 hours.

(1) Remove the cover to get across[4] to the bearings for the balance drive.

(2) Fill a grease pump with the grease of the specified type.

grease pump
润滑脂泵

(3) Press a small portion of grease into each of the 4 bearings (as shown in Fig. 7-2).

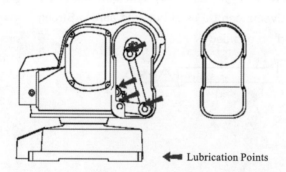

← Lubrication Points

Fig. 7-2 Lubrication Points 1

lubrication point
润滑点

(4) Check the sealing ring for damage, adjust it properly, and reinstall it.

(5) Repeat the operation for bearings for axis 2 on the opposite side[5].

3. Part 3 Lubrication of Balance Spring Rods

(1) Move the vertical arm (axis 2) to the forward or backward position, and the horizontal arm (axis 3) down to expose the balance

vertical arm
垂直臂

spring rods to be lubricated.

(2) Remove the spring cover for axis 2 on the left side at the rear of the robot base.

(3) Use a brush or similar tool, and add a portion of the specified type of grease to the balance spring rod (as shown in Fig. 7-3).

horizontal arm
水平臂
robot base
机器人底座

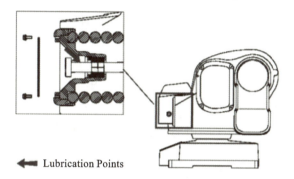

Fig. 7-3　Lubrication Points 2

(4) Re-install the spring cover.

(5) Repeat the operation for the balance spring rod (axis 3) on the right side.

4. Part 4　Lubrication of Hollow Wrists

(1) Using a small screwdriver, open seals as shown in the lower right in Fig. 7-4. Add a little lubricant of the specified type at 5 or 6 different positions around the seal.

screwdriver
螺丝刀

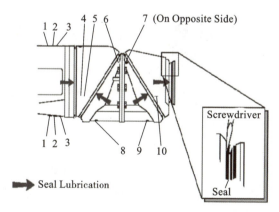

Fig. 7-4　Seal Lubrication

(2) Repeat the operation for all 4 seals at the sides marked with arrows in the illustration.

(3) Fill a grease pump with[6] the grease of the specified type.

Note: when lubricating the wrist, a few drops of lubricant is sufficient at each grease nipple. Do not insert a larger portion of lubricant, as this may damage the wrist seals and internal sleeves.

grease nipple
润滑脂头
internal sleeve
内部套筒

(4) Lubricate grease nipples 1-10, one by one, by pressing a very small portion of grease into each nipple.

5. Part 5　Check Oil Level in Gearboxes for Axis 1-6

The gearboxes are lubricated for life with oil which corresponds to 40 000 hours in operation if the robot is operating in an environment temperature below 40 degrees.

To check the oil level, open the filling plugs and check that the oil level is up to the filling hole(as shown in Fig. 7-5). If not, top up with the specified oil type. For specification of oil type and filling instructions, refer to the section of gearbox oil change in the repair manual.

filling hole
充油孔

upper bevel gear
上锥齿轮

filling plug
充油塞

gearbox
齿轮箱

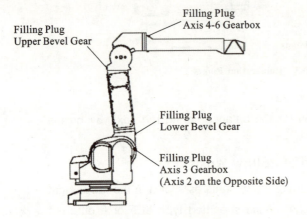

Fig. 7-5　Gearbox Lubrication

 Tips

1. ABB: Asea Brown Boveri 公司, 简称 ABB 公司

ABB 公司是世界 500 强企业, 公司总部位于瑞士苏黎世。ABB 公司由两个具有 100 多年历史的国际性企业, 即瑞典的阿西亚公司(ASEA)和瑞士的布朗勃法瑞公司(BBC Brown Boveri)合并而成。ABB 公司是电力和自动化技术领域的领先者。ABB 公司业务遍布全球 100 多个国家和地区。

 New Words and Phrases

manual [ˈmænjuəl] *n.* 使用手册, 风琴键盘　*adj.* 手动的, 手工的, 手控的

bearing [ˈbeərɪŋ] n. 轴承,定向,方位
manipulator [məˈnɪpjuleɪtə(r)] n. 机械手,调制器,操作者
lubricate [ˈluːbrɪkeɪt] v. 润滑,给……加润滑油,促进
regularly [ˈregjələli] adv. 定期地,有规律地,频繁地,经常地
contaminated [kənˈtæməneɪtɪd] adj. 受污染的
hollow [ˈhɒləʊ] adj. 空的,凹陷的,虚伪的
hollow wrist 空心腕关节
lint [lɪnt] n. 棉绒,毛絮,线头
lint-free cloth 无绒布
excessive [ɪkˈsesɪv] adj. 过度的,过多的
grease [griːs] n. 油脂,润滑油
portion [ˈpɔːʃ(ə)n] n. 一份,一部分 v. 分配
seal [siːl] n. 密封,密封状态,水封
vertical [ˈvɜːtɪk(ə)l] adj. 垂直的,直立的,纵向的
vertical arm 垂直臂
sleeve [sliːv] n. (机器的)套筒,套管,袖子
internal sleeve 内部套筒
plug [plʌg] n. (电)插头,塞子 v. 堵,塞,补足
filling plug 充油塞
bevel [ˈbevl] n. 斜角,斜面,[测]斜角规
upper bevel gear 上锥齿轮
specification [ˌspesɪfɪˈkeɪʃ(ə)n] n. 规范,明确说明,详述
balance [ˈbæləns] n. 平衡,均衡,均势
balance drive bearing 平衡传动轴承
spring [sprɪŋ] n. 弹簧,发条,春天,泉水
balance spring rod 平衡弹簧杆
arm assembly 手臂组件
strong solvent 强溶剂
vaseline [ˈvæsəliːn] n. 凡士林
service personnel 维修人员
purge [ˈpɜːrdʒ] v. 清洗,净化
purging unit 清洁装置
lubrication point 润滑点
grease pump 润滑脂泵
horizontal [ˌhɒrɪˈzɒnt(ə)l] adj. 水平的,统一的
horizontal arm 水平臂
robot base 机器人底座
screwdriver [ˈskruːdraɪvə(r)] n. 螺丝刀

nipple ['nɪpl] n. 乳头,奶嘴
grease nipple 润滑脂头
gearbox ['ɡɪəbɒks] n. 齿轮箱
filling hole 充油孔

Notes

1. with care 小心地

例句:Solvents can be used but with care.

可以使用溶剂,但要谨慎。

例句:Treat your keyboard with care then it can last for years.

爱惜你的键盘,这样就可以使用很多年。

2. add to 加入,加到,增加

例句:After the wrist is cleaned, a small amount of vaseline should be added to the surface of the wrist.

手腕清洁后,应在手腕表面加少量凡士林。

例句:I have nothing to add to my earlier statement.

我对我早先说的话没有什么补充的。

3. call for 联系,需要,接

例句:For oil leaks, if serious oil leakage is found, call for service personnel.

对于漏油,如果发现严重漏油,请联系服务人员。

例句:I'll call for you at seven o'clock.

我7点钟去接你。

4. get across 通过,使……被理解

例句:Remove the cover to get across to the bearings for the balance drive.

拆下盖子,穿过平衡传动的轴承。

例句:The bridge was destroyed so we couldn't get across the river.

大桥已经毁坏,我们无法过河了。

5. on the opposite side 在另一边,在对面

例句:Repeat the operation for bearings for axis 2 on the opposite side.

对另一侧2轴的轴承重复此操作。

例句:We live further down on the opposite side of the road.

我们住在马路对面再远一点的地方。

6. fill...with... 充满,填塞,充实

例句:Fill a grease pump with the grease of the specified type.

向润滑脂泵中注入指定类型的润滑脂。

例句:It takes the inflatable chair more than five minutes to fill with air.
至少需要 5 分钟才能将充气椅充满空气。

Exercises

I. Write True or False beside the following statements about the text.

1. _____ The manipulator base and arm assembly should be cleaned regularly.
2. _____ Strong solvents such as acetone could be used.
3. _____ After the wrist is cleaned, a small amount of vaseline should be added to the surface of the wrist.
4. _____ For oil leaks, if serious oil leakage is found, call for service personnel.
5. _____ To check the oil level, open the filling plugs and check that the oil level is up to the filling hole.

II. Answer the following questions in English according to the text.

1. What are the general steps for the maintenance of manipulators?
2. How do you lubricate the balance drive bearings?
3. How do you lubricate the hollow wrist?

III. Read the text again and fill in the blanks in the following sentences orally.

1. Clean the external contaminated paint parts with a _____ _____ after each shift and during the break. After cleaning, carefully dry the components with _____ _____.
2. The wrist must be cleaned as often as required. And take care to avoid the accumulation of dirt and _____ _____. Use lint-free cloth for cleaning.
3. Fill a grease pump with the _____ of the specified type.
4. Move the _____ _____ (axis 2) to the forward or backward position, and the _____ _____ (axis 3) down to expose the balance spring rods to be lubricated.
5. To check the oil level, open the filling plugs and check that the oil level is up to the _____ _____.

IV. Translation.

The wrist must be cleaned as often as required. And take care to avoid the accumulation of dirt and dust particles. Use lint-free cloth for cleaning. After the wrist is cleaned, a small amount of vaseline should be added to the surface of the wrist. This will simplify the next cleaning.

Lesson 8　The Maintenance of the Manipulator Control Cabinet

Table 8-1 lists control cabinet maintenance. This manual is used for "quick reference" of the ABB robot manual. Please read the ABB robot manual in detail.

Table 8-1　Control Cabinet Maintenance

No.	Activity and Reference	Interval
1	Clean / Replace Filter (Part 6)	Daily
2	Changing the Measuring System Battery (Part 7)	3 000 hours
3	Updating Revolution Counters (Part 8)	3 000 hours

Margin Note

revolution counter
转数计数器

1. Part 6　Clean / Replace Filter

(1) Locate the filter on the back of the cabinet.

(2) Lift the filter holder up and out.

Note: on the compact cabinet, a string is attached to[①] the filter and the top of the cabinet. This string can be used to pull the filter out.

(3) Remove the old filter from the filter holder.

(4) Insert the new filter in the filter holder.

As an alternative to replacing the filter, the filter can be cleaned. Clean the filter three or four times with washing-up detergent in water at 30-40 ℃. The filter should not be wrung out[②] but should be allowed to[③] dry on a flat surface.

Alternatively, the filter can be blown clean with compressed air from the clean-air side.

(5) Slide the filter holder with the new filter in and down into position.

Cabinet front and rear view are shown in Fig. 8-1. Cabinet rear and side view are shown in Fig. 8-2.

filter holder
过滤器座

compressed air
压缩空气

2. Part 7　Changing the Measuring System Battery

(1) Open the front door of the control cabinet and swing the rotating unit aside to get access to[④] the backup battery.

(2) Disconnect the battery connector.

(3) Remove the screw and battery clamp. Remove the old battery.

(4) Install the new battery with the same clamp and screw. And place the new battery in the same way. Then check that the battery clamp is

control cabinet
控制柜
backup battery
备用电池
battery connector
电池连接器
battery clamp
电池夹

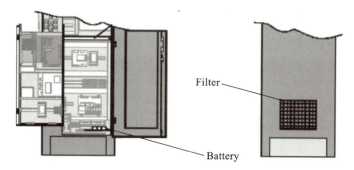

Fig. 8-1 Cabinet Front and Rear View

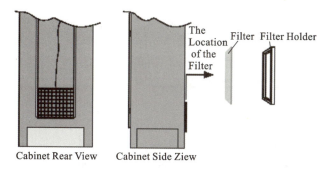

Cabinet Rear View   Cabinet Side Ziew

Fig. 8-2 Cabinet Rear and Side View

secure.

(5) Connect the battery connector.

(6) Swing the turning device back into position and close the door of the cabinet.

(7) Perform revolution counter update as described in Part 8 of this manual.

(8) Charge the battery. An uncharged battery takes 36 hours to recharge. The mains supply must be switched on[5] during this time.

Changing the measuring system battery is shown in Fig. 8-3.

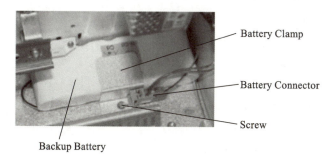

Fig. 8-3 Changing the Measuring System Battery

3. Part 8 Updating Revolution Counters

When there is a problem with the battery or some devices, a message may be displayed telling you that the "revolution counter" is not updated. The message appears in form of⑥ an error message on the teach pendant (as shown in Fig. 8-4). If you receive such a message, the revolution counter of the manipulator must be updated with the calibration marks on the manipulator.

teach pendant
示教器

calibration mark
校准标记

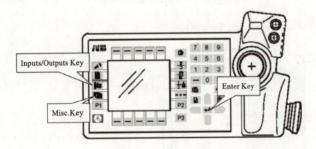

Fig. 8-4 Teach Pendant

(1) Press the Misc. key ![icon] on the teach pendant.

(2) Select the "Service" in the dialog box on the display (as shown in Fig. 8-5).

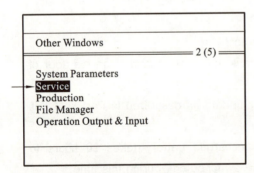

Fig. 8-5 Display of the Teach Pendant

(3) Press the Enter key ↵ .

(4) Select "3 calibration" in the "View" drop-down menu (as shown in Fig. 8-6) and press the Enter key ↵ .

drop-down menu
下拉菜单

(5) Select "IRB". IRB represents Axis 1 to Axis 6.

Choose "Calib" and "1 Rev. counter update ..." (as shown in Fig. 8-7) and press the Enter key ↵ .

(6) The window (as shown in Fig. 8-8) will appear.

(7) Place the robot in the calibration position (Axis 1 to Axis 6). Illustrations (as shown in Fig. 8-9, Fig. 8-10, and Fig. 8-11) show the location of the calibration marks.

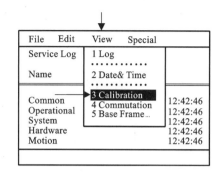

Fig. 8-6　Choose "View"

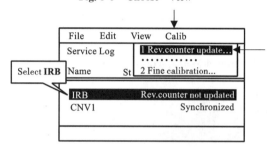

Fig. 8-7　Choose "Calib"

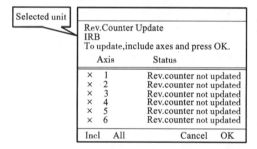

Fig. 8-8　The Window Shown on the Display

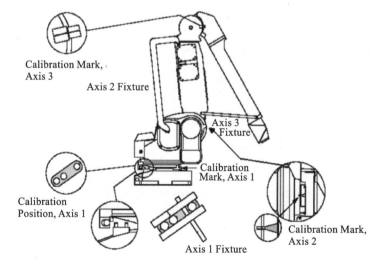

Fig. 8-9　The Location of the Calibration Marks, Axis 1 to Axis 3

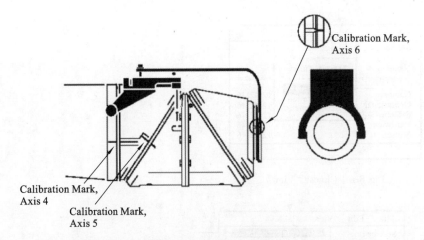

Fig. 8-10　The Location of the Calibration Marks, Axis 4 to Axis 6

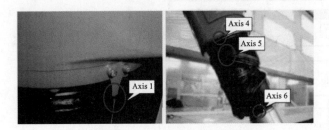

Fig. 8-11　The Location of the Axis

(8) Select the required axis for which the revolution counter is to be updated. Press the function key "Incl"(as shown in Fig. 8-12). The selected axis will be marked with an "X".

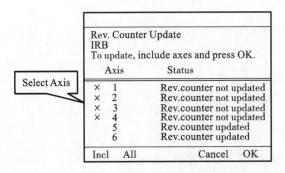

Fig. 8-12　Select Axis

(9) To proceed "revolution counter update", we need to confirm. Start updating the revolution counter by pressing "OK"(as shown in Fig. 8-13).

(10) After updating "Rev. Counter", check the calibration position again.

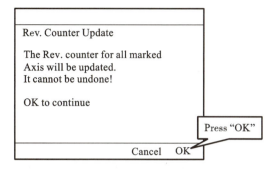

Fig. 8-13　Update Revolution Counter

 New Words and Phrases

cabinet [ˈkæbɪnət] n. 机箱,储藏柜,陈列柜
control cabinet 控制柜
filter [ˈfɪltə(r)] n. 过滤器,滤声器,滤波器
filter holder 过滤器座
backup battery 备用电池
battery connector 电池连接器
revolution [ˌrevəˈluːʃn] n. 旋转,旋转一周,革命
revolution counter 转数计数器
locate [ləʊˈkeɪt] v. 确定……的位置,设立,建立
string [strɪŋ] n. 线,一连串,一系列(事件)
remove [rɪˈmuːv] v. 移开,除去,废除
insert [ɪnˈsɜːt] v. 插入,嵌入,(在文章中)添加
detergent [dɪˈtɜːdʒənt] n. 洗涤剂,去污剂
compressed [kəmˈprest] adj. (被)压缩的,扁的
compressed air 压缩空气
disconnect [ˌdɪskəˈnekt] v. 切断,使分离
clamp [klæmp] n. 夹具,钳位电路
battery clamp 电池夹
charge [tʃɑːdʒ] v. 充电,使……承担责任
calibration [ˌkælɪˈbreɪʃn] n. 标定,(测量器上的)刻度
calibration mark 校准标记
illustration [ˌɪləˈstreɪʃ(ə)n] n. 插图,说明,实例
teach pendant 示教器

Notes

1. be attached to 将某物系在，附属于

例句：On the compact cabinet, a string is attached to the filter and the top of the cabinet.

在紧凑型控制柜上，过滤器和控制柜顶部之间有一根绳子。

例句：For example, annotations might be attached to a span of text in a document.

例如，注释可能被附加到文档中的一段文本上。

2. wring out 拧干，扭干

例句：The filter should not be wrung out but should be allowed to dry on a flat surface.
过滤器不能拧干，而应该将其放在一个平面上晾干。

例句：He turned away to wring out the wet shirt.
他转身去拧干那件湿衬衫。

3. be allowed to 被允许做某事

例句：The filter should not be wrung out but should be allowed to dry on a flat surface.
过滤器不能拧干，而应该将其放在一个平面上晾干。

例句：She begged that she should be allowed to go.
她请求让她离开。

4. get access to 获得，接近

例句：Open the front door of the control cabinet and swing the rotating unit aside to get access to the backup battery.
打开控制柜前门，将旋转装置摆到一边，直至接触到备用电池。

例句：You need a password to get access to the computer system.
使用这个计算机系统需要口令。

5. switch on 接通，开启

例句：The mains supply must be switched on during this time.
在这段时间内，必须接通电源。

例句：Switch on, push in the disk and there you are!
你打开开关，把磁盘推进去就行了！

6. in form of 以……的形式

例句：The message appears in form of an error message on the teach pendant.
该消息以错误消息的形式出现在示教器上。

例句：They received a benefit in form of a tax reduction.
他们以减税的方式获得了收益。

Exercises

Ⅰ. Write True or False beside the following statements about the text.

1. _____ On the compact cabinet, a string is attached to the filter and the top of the cabinet. This string may be used to pull the filter out.

2. _____ Alternatively, the filter can be blown clean with compressed air from the clean-air side.

3. _____ Charge the battery. An uncharged battery takes 36 hours to recharge. The mains supply must be switched off during this time.

4. _____ When there is a problem with the battery or some devices, a message may be displayed telling you that the "revolution counter" is not updated.

5. _____ Place the robot in the calibration position(Axis 1 to Axis 6).

Ⅱ. Answer the following questions in English according to the text.

1. How do you clean and replace the filter according to the ABB robot manual?

2. When a battery or some equipment has a problem, what are the general steps to update the revolution counters?

3. How can the robot be placed in the calibration position? Illustrate it according to the figures.

Ⅲ. Read the text again and fill in the blanks in the following sentences orally.

1. As an alternative to replacing the filter, the filter may be cleaned. Clean the filter three or four times in 30-40 °C water with _____ _____.

2. Open the _____ _____ of the control cabinet and swing the rotating unit aside to get access to the _____ _____.

3. Install the new battery with the same _____ and _____. And place the new battery in the same way. Then check that the _____ _____ is secure.

4. Charge the _____. An uncharged battery takes 36 hours to recharge. The mains supply must be _____ _____ during this time.

5. Select the required axis for which the _____ _____ is to be updated. Press the function key "Incl". The _____ _____ will be marked with an "X".

Ⅳ. Translation.

When there is a problem with the battery or some devices, a message may be displayed telling you that the "revolution counter" is not updated. The message appears in form of an error message on the teach pendant(as shown in Fig. 8-4). If you receive such a message, the revolution counter of the manipulator must be updated with the calibration marks on the manipulator.

Lesson 9　Autonomous Mobile Robotics

An autonomous mobile robot(AMR) is a mechanical device that can perform its work automatically. It is an intelligent robot with self-organization, independent operation, and independent planning, and it can work in complex environments. Autonomous mobile robots vary greatly with[①] their application environment and moving mode. Their common basic technology integrates computer technology, information technology, communication technology, microelectronics, sensor technology, mobile technology, operational control technology, robotics, and artificial intelligence. An autonomous mobile robot is a robot system composed of[②] sensors, remote operators, and mobile carriers of automatic control devices. It is equipped with[③] smart devices such as visual sensors, auditory sensors, and tactile sensors, which are equivalent to[④] human eyes, ears, and skin.

Through the application of artificial intelligence technology, the mobile robot systems can autonomously complete the work of reasoning, planning, and controlling in complex environments. As shown in Fig. 9-1, autonomous mobile robots can replace human work in dangerous and harsh places (e.g. toxic, radioactive, etc.) or places that are unsuitable for human work (e.g. space, underwater, etc.). They have greater mobility and flexibility than ordinary robots, representing the highest achievement of mechatronics technology nowadays.

Margin Note
autonomous
mobile robot
自主移动机器人

Fig. 9-1　Autonomous Mobile Robots Replace Human Work in Dangerous Places

According to the way of moving, the intelligent mobile robots can be divided into wheeled mobile robots (as shown in Fig. 9-2), walking

wheeled
mobile robot
轮式移动机器人

mobile robots (single-legged, double-legged, and multi-legged), tracked mobile robots (as shown in Fig. 9-3), crawling robots (as shown in Fig. 9-4), peristaltic robots, swimming robots and other types. According to the working environment, they can be divided into indoor mobile robots and outdoor mobile robots. According to the control architecture, they can be divided into functional (horizontal type) structure robots, behavior (vertical type) structure robots, and hybrid robots. According to functions and uses, they can be divided into medical robots, military robots, assistive robots (as shown in Fig. 9-5), cleaning robots (as shown in Fig. 9-6), etc. The control mode of mobile robots develops from remote control and monitoring to autonomous control. Mobile robots have integrated the application of machine vision, problem-solving, expert systems, and other artificial intelligence technologies, and constantly improved into a higher level of autonomous mobile robots.

tracked mobile robot
履带式移动机器人
crawling robot
爬行机器人
peristaltic robot
蠕动机器人

hybrid robot
混合机器人

machine vision
机器视觉

Fig. 9-2　Wheeled Mobile Robot

Fig. 9-3　Tracked Mobile Robot

Fig. 9-4　Crawling Robot

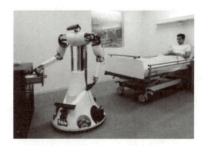

Fig. 9-5　Assistive Robot

Fig. 9-6　Swimming Pool Cleaning Robot

At present, the most advanced autonomous mobile robots can

autonomously optimize workflow and save staff resources, thereby improving productivity and reducing costs. With advanced navigation software instead of cables, magnets, or QR codes, robots can use positioning technology to orient themselves within the operating environment and navigate to find the best path to their destination by downloading a CAD file of the facility to the robot or using a laser scanner to create a map. At the same time, a mobile robot automatically avoids dynamic obstacles or timely safety stops according to the sensor input. A 3D front camera is mounted on the robot, which improves navigation performance and docking accuracy. The mobile robot can carry objects up to 500 kg to 1 000 kg, such as boxes, shelves, lifting devices, conveyor belts, and even cooperative robot arms, and move in a highly flexible and autonomous way. By replacing the module, the robot can be reconfigured for different tasks.

laser scanner
激光扫描器

conveyor belt
传送带

Since the batteries can be replaced quickly, autonomous mobile robots can work all day long, helping to boost productivity. They can be operated on a smartphone, a tablet, or a computer through an intuitive robot interface and can be programmed based on previous experiences.

 Tips

1. microelectronics 微电子技术

微电子技术是随着集成电路,尤其是超大规模集成电路而发展起来的一门新技术。微电子技术包括系统电路设计、器件物理、工艺技术、材料制备、自动测试以及封装、组装等一系列专门技术,是微电子学中的各项工艺技术的总和。微电子技术可在纳米级超小的区域内通过固体内的微观电子运动来实现信息的处理与传递,并且有着很好的集成性。其核心在于集成电路,它是在各类半导体器件不断发展过程中形成的。在信息化时代,微电子技术对人类生产、生活都带来了极大的影响。

2. QR Code:Quick Response Code QR 码

QR 码是 Denso 公司于 1994 年 9 月研制的一种矩阵二维码符号,它呈正方形,只有黑白两色。在 4 个角落中的其中 3 个,印有较小、像"回"字的正方形图案,它们是帮助解码软件定位的图案。使用者无论以何种角度扫描,资料都可被正确读取。与条形码相比,QR 码存储的信息更丰富、读取更快速、可靠性高、可表示汉字及图像、保密防伪性强。超高速识读的特性使它能够广泛应用于工业自动化生产线管理等领域。

New Words and Phrases

autonomous mobile robot 自主移动机器人
mechanical [mə'kænɪkl] *adj*. 机械的,机械驱动的,呆头呆脑的,机械学的
microelectronics [ˌmaɪkrəʊɪlek'trɒnɪks] *n*. 微电子技术,微电子学
auditory ['ɔːdətri] *adj*. 听觉的,听的
tactile ['tæktaɪl] *adj*. 触觉的,有触觉的,能触知的
equivalent [ɪ'kwɪvələnt] *adj*. 相等的,相同的 *n*. 相等的东西,等量
toxic ['tɒksɪk] *adj*. 有毒的,引起中毒的,卑鄙无耻的
radioactive [ˌreɪdiəʊ'æktɪv] *adj*. 放射性的,有辐射的
mechatronics [ˌmekə'trɒnɪks] *n*. 机械电子学
wheeled [wiːld] *adj*. 轮式的,有轮的
wheeled mobile robot 轮式移动机器人
tracked mobile robot 履带式移动机器人
crawl [krɔːl] *v*. 爬行,匍匐前进,缓慢前进
crawling robot 爬行机器人
peristaltic [ˌperɪ'stæltɪk] *adj*. 蠕动的,蠕动引起的
peristaltic robot 蠕动机器人
hybrid ['haɪbrɪd] *adj*. 混合的 *n*. 合成物
hybrid robot 混合机器人
machine vision 机器视觉
QR code QR 码
laser scanner 激光扫描器
obstacle ['ɒbstəkl] *n*. 障碍,阻碍,绊脚石,障碍栅栏
dock [dɒk] *v*. 对接,(船舶)进坞
conveyor [kən'veɪə(r)] *n*. 传输带,运送者,传播者
conveyor belt 传送带
cooperative [kəʊ'ɒpərətɪv] *adj*. 协作的,同心协力的,配合的
module ['mɒdjuːl] *n*. 模块,功能块,程序块,组件,配件,舱,单元
reconfigure [ˌriːkən'fɪɡə(r)] *v*. 重新配置(计算机设备等),重新设定(程序等)
intuitive [ɪn'tjuːɪtɪv] *adj*. 直观的,直觉的,凭直觉得到的,有直觉力的,易懂的

 Notes

1. vary with 随……变化

例句:Autonomous mobile robots <u>vary</u> greatly <u>with</u> their application environment and moving mode.

自主移动机器人随其应用环境和移动方式的不同有很大差别。

例句:The menu varies with the season.

菜单随季节而变动。

2. be composed of 由……组成(或构成)

例句:An autonomous mobile robot is a robot system composed of sensors, remote operators, and mobile carriers of automatic control devices.

自主移动机器人是一种由传感器、遥控操作器和自动控制装置的移动载体组成的机器人系统。

例句:The committee is composed mainly of lawyers.

委员会主要由律师组成。

3. be equipped with 配有(某种装置)

例句:It is equipped with smart devices such as visual sensors, auditory sensors, and tactile sensors, which are equivalent to human eyes, ears, and skin.

它配有与人类的眼睛、耳朵和皮肤相当的视觉传感器、听觉传感器和触觉传感器等智能装置。

例句:You will be equipped with all the weapons you need.

你将拥有你所需要的所有武器装备。

4. be equivalent to 与……相当的,与……相等的

例句:It is equipped with smart devices such as visual sensors, auditory sensors, and tactile sensors, which are equivalent to human eyes, ears, and skin.

它配有与人类的眼睛、耳朵和皮肤相当的视觉传感器、听觉传感器和触觉传感器等智能装置。

例句:Eight kilometers is roughly equivalent to five miles.

八公里约等于五英里。

Exercises

Ⅰ. Write True or False beside the following statements about the text.

1. _____ An autonomous mobile robot is a robot system composed of sensors, remote operators, and mobile carriers of automatic control devices.

2. _____ Autonomous mobile robots have greater mobility and flexibility than ordinary robots.

3. _____ According to functions and uses, autonomous mobile robots can be divided into medical robots, military robots, assistive robots, cleaning robots, etc.

4. _____ Mobile robots can't carry heavy loads.

5. _____ Autonomous mobile robots can't work all day long because of battery shortage.

II. Answer the following questions in English according to the text.

1. What does AMR refer to?
2. How to categorize AMRs according to the text?
3. Summarize the key advantages of autonomous mobile robots.

III. Read the text again and fill in the blanks in the following sentences orally.

1. An AMR is a robot system composed of _____, remote operators, and mobile carriers of _____ control devices. It is _____ with smart devices such as _____ sensors, _____ sensors, and _____ sensors, which are _____ to human eyes, ears, and skin.

2. According to the way of moving, the intelligent mobile robots can be divided into _____ mobile robots, _____ mobile robots, _____ mobile robots, _____ robots, _____ robots, swimming robots, and other types.

3. The control mode of mobile robots develops from _____ _____ and _____ to _____ _____. Mobile robots have _____ the application of _____ _____, problem-solving, _____ systems and other _____ _____ technologies, and constantly improved into a higher level of autonomous mobile robots.

4. The mobile robot can carry objects up to 500 kg to 1 000 kg, such as boxes, shelves, _____ _____, _____ _____, and even _____ robot arms, and move in a highly _____ and _____ way.

5. Autonomous mobile robots can be _____ on a smartphone, a tablet, or a computer through an _____ _____ _____ and can be _____ based on previous experiences.

IV. Translation.

At present, the most advanced autonomous mobile robots can autonomously optimize workflow and save staff resources, thereby improving productivity and reducing costs. With advanced navigation software instead of cables, magnets, or QR codes, robots can use positioning technology to orient themselves within the operating environment and navigate to find the best path to their destination by downloading a CAD file of the facility to the robot or using a laser scanner to create a map.

Lesson 10　Basic Operation Training Tasks of Industrial Robots

New employees should receive industrial robot operation training before they start to work formally, to understand the operation precautions and operation safety specifications, so as to ensure that

Margin Note

they can undertake the job tasks smoothly.

1. Precautions for Industrial Robot Operation Training

(1) Operate the robot through the operation panel and the teaching pendant. Do not touch the robot body and the interior of the electrical cabinet.

electrical cabinet
电器柜

(2) When checking the teaching pendant or other structures of the robot after startup, press the emergency stop button to prevent the robot from[①] misoperation.

(3) In continuous motion mode, the override shall not exceed 20%.

(4) When operating the robot, carefully estimate the trajectory of the robot before the operation, so as to avoid robot damage or personal injury caused by misoperation.

(5) When the end of the robot is close to the target point, adjust the override or select the incremental mode for operation.

(6) When performing the manual operation, keep away from the robot as far as possible. Do not stay in the fence unless necessary.

(7) Do not run the program by yourself without the presence of the instructor.

2. Safety Specification for the Operation of Industrial Robots

(1) The personnel who teach, debug, and maintain the robot must receive safety and operation training in advance.

(2) Dress code for robot operation:

① Do not wear gloves when operating the robot.

② Wear work clothes and a safety helmet when operating the robot.

③ Do not wear jewelry, such as earrings or pendants.

(3) Do not lean on the robot, robot control cabinet, or other control cabinets.

(4) During operation, do not allow non-staff to touch the control cabinet, as shown in Fig. 10-1.

Fig. 10-1 Do Not Lean on the Robot Control Cabinet and do not Allow Non-staff to Touch the Control Cabinet

(5) When installing tools on the robot, be sure to cut off the power supply first. If the power supply is connected during installation, an electric shock may be caused, or abnormal movement of the robot may occur, resulting in injury. In addition, the weight of the tool must not exceed the allowable range of the robot.

electric shock
电击

(6) Before operating the robot, first press the emergency stop key on the teaching pendant to check whether the servo is powered off and confirm that the power supply has been turned off. In case of emergency, if the robot cannot be stopped, personal injury or equipment damage will occur.

(7) Before performing the following operations, confirm that there is no person within the action range of the robot and that the personnel operates within the safe range.

①The control cabinet is powered on.

②Programming at the teaching pendant to move the robot.

③The debugging program is running.

④The program is running automatically.

(8) If you accidentally enter the movable range of the robot, it may cause personal injury. In case of② abnormal operation, please directly press the emergency stop button. The emergency stop button is on the upper right side of the teaching pendant.

(9) Before teaching the robot, the following items shall be checked. If there is any abnormal condition, repair it immediately or take necessary measures.

①Check for problems with robot movement.

②Check whether the insulation protection cover of the external cable is damaged.

insulation protection cover
绝缘保护罩

(10) The teaching pendant must be returned to its original position after use. If the teaching programmer is inadvertently placed on the robot, fixture, or floor, when the robot is working, it will touch the teaching pendant on the robot or tool, which may cause personal injury or equipment damage.

Fig. 10-2　Never Force a Robot's Shaft to Move

(11) Never force a robot's shaft to move, as shown in Fig. 10-2.

New Words and Phrases

panel ['pænl] n. 面板
pendant ['pendənt] n. 器,吊坠,(项链上的)垂饰
exceed [ɪk'siːd] v. 超过
estimate ['estɪmət,'estɪmeɪt] v. 估计
trajectory [trə'dʒektəri] n. 轨迹
adjust [ə'dʒʌst] v. 调整,调节
debug ['diːbʌg] v. 排错,调试
helmet ['helmɪt] n. 头盔,防护帽
abnormal [æb'nɔːml] adj. 不正常的,反常的
allowable [ə'lauəbl] adj. 允许的,承认的,容许的
accidentally [ˌæksɪ'dentəli] adv. 意外地,偶然地
external [ɪk'stɜːnl] adj. 外部的,外面的
original [ə'rɪdʒənl] adj. 起初的,原来的
inadvertently [ˌɪnəd'vɜːtəntli] adv. 无意地,不经意地
electrical cabinet 电器柜
electric shock 电击
insulation [ˌɪnsju'leɪʃ(ə)n] n. 隔热,绝缘
insulation protection cover 绝缘保护罩

 Notes

1. prevent...from 阻止

例句:When checking the teaching pendant or other structures of the robot after startup,press the emergency stop button to prevent the robot from misoperation.
开机后查看示教器或机器人其他结构时,按下急停开关,以防机器人误动作。
例句:This can prevent air from flowing freely to the lungs.
这可以防止空气随意流入肺部。

2. in case of 如果发生……;若在……情况下,万一

例句:In case of abnormal operation,please directly press the emergency stop button.
若异常,请直接按急停按钮。
例句:In case of emergency,break the glass and press the button.
遇到紧急情况时,击碎玻璃并按下按钮。

Exercises

Ⅰ. Write True or False beside the following statements about the text.

1. _____ Do not touch the robot body and the interior of the electrical cabinet.

2. _____ In continuous motion mode, the override shall not exceed 10%.

3. _____ The personnel who teach, debug and maintain the robot must receive safety and operation training in advance.

4. _____ Wear gloves when operating the robot.

5. _____ When installing tools on the robot, be sure to cut off the power supply first.

Ⅱ. Answer the following questions in English according to the text.

1. What will occur if the power supply is connected during installation?

2. What should be done before operating the robot?

3. Where is the emergency stop button?

Ⅲ. Read the text again and fill in the blanks in the following sentences orally.

1. Operate the robot through the _____ _____ and the _____ _____.

2. When operating the robot, carefully _____ the trajectory of the robot before the operation.

3. Do not run the program by yourself without the presence of the _____.

4. In case of emergency, if the robot cannot be stopped, _____ _____ or equipment damage will occur.

5. The teaching pendant must be returned to its _____ position after use.

Ⅳ. Translation.

When installing tools on the robot, be sure to cut off the power supply first. If the power supply is connected during installation, an electric shock may be caused, or abnormal movement of the robot may occur, resulting in injury. In addition, the weight of the tool must not exceed the allowable range of the robot.

 ## Reading Material 5 Installation of the Manipulator

To ensure production safety, be sure to install safeguards. Read the following information and guidance carefully to prevent personal accidents and equipment damage.

Margin Note

1. Warning

(1) Install the safeguards. Failure to observe this warning may result in injury or damage.

(2) Install the manipulator in a location where the tool or the workpiece held by its fully extended arm will not reach the wall, safeguarding, or controller. Failure to observe this warning may result in injury or damage.

(3) Do not start the manipulator or even turn on the power before it is firmly anchored. The manipulator may overturn and cause injury or damage.

(4) When mounting the manipulator on the ceiling or wall, the base section must have sufficient strength and rigidity to support the weight of the manipulator. Also, it is necessary to consider countermeasures to prevent the manipulator from falling. Failure to observe these warnings may result in injury or damage.

2. Caution

(1) Do not install or operate a manipulator which is damaged or lacks parts. Failure to observe this caution may cause injury or damage.

(2) Before turning on the power, check to be sure that the shipping bolts and brackets explained in "Transporting Position" are removed. Failure to observe this caution may result in damage to the driving parts.

The user of a manipulator or robot system shall ensure that safeguards are provided and used in accordance with this standard. The means and degree of safeguards, including any redundancies, shall correspond directly to the type and level of hazard presented by the robot system consistent with the robot application. Safeguarding may include but not be limited to safeguarding devices, barriers, interlock barriers, perimeter guarding, awareness barriers, and awareness signals.

3. Mounting Procedures for Manipulator Base

The manipulator should be firmly mounted on a baseplate or foundation strong enough to support the manipulator and withstand repulsion forces during acceleration and deceleration. Construct a solid foundation with the appropriate thickness to withstand the maximum repulsion forces of the manipulator, as shown in Table R5-1 and Table R5-2.

Table R5-1 The Maximum Repulsion Forces of the Manipulator at Emergence Stop

Horizontal rotating maximum torque (S-axis moving direction)	700 N·m
Vertical rotating maximum torque (LU-axis moving direction)	700 N·m

Table R5-2 Endurance Torque in Operation

Endurance torque in horizontal operation (S-axis moving direction)	220 N·m
Endurance torque in vertical operation (LU-axis moving direction)	270 N·m

4. Mounting Example

For the first process, anchor the baseplate firmly to the ground. The baseplate should be rugged and durable to prevent shifting of the manipulator or the mounting fixture. It is recommended to prepare a baseplate of 30 mm or more in thickness, and anchor bolts of M10 or larger in size.

anchor bolt
地脚螺栓

The manipulator base (as shown in Fig. R5-1) is tapped for four mounting holes: securely fix the manipulator base to the baseplate with four hexagon socket head cap screws M10 (recommended length: 35 mm). Next, fix the manipulator base to the baseplate. Tighten the hexagon socket head cap screws and anchor bolts firmly so that they will not work loose during the operation.

hexagon socket
head cap screw
内六角圆柱头螺钉

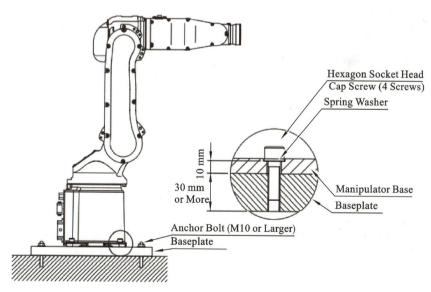

Fig. R5-1 **Mounting the Manipulator on Baseplate**

New Words and Phrases

anchor [ˈæŋkə(r)] v. 使固定
overturn [ˌəʊvəˈtɜːn] v. 倾倒,倾覆,翻掉
rigidity [rɪˈdʒɪdətɪ] n. 刚性,强直,严格
countermeasure [ˈkaʊntəmeʒə(r)] n. 对策,对抗手段,反措施
in accordance with 符合,依照,和……一致
redundancy [rɪˈdʌndənsi] n. 多余,累赘
consistent with 与某事物并存(一致)
interlock [ˌɪntəˈlɒk] v. (使)连锁
perimeter [pəˈrɪmɪtə(r)] n. 外缘,边缘
baseplate [beɪspleɪt] n. 底板,撑板,底盘
torque [tɔːk] n. (使机器等旋转的)转矩
endurance [ɪnˈdjʊərəns] n. 耐久力,忍耐力
repulsion [rɪˈpʌlʃn] n. 排斥力,斥力
rugged [ˈrʌgɪd] adj. 结实的,耐用的
bolt [bəʊlt] n. 螺栓
anchor bolt 地脚螺栓
durable [ˈdjʊərəbl] adj. 耐用的,持久的
hexagon [ˈheksəgən] n. 六边形
hexagon socket head cap screw 内六角圆柱头螺钉
socket [ˈsɒkɪt] n. 承窝,承槽,插孔
tighten [ˈtaɪtn] v. (使)变紧,更加牢固

Reading Material 6   Robot Parts List

1. S-axis Unit

S-axis Unit is shown in Fig. R6-1 and Table R6-1 respectively.

Unit II Industrial Robot Technology | 79

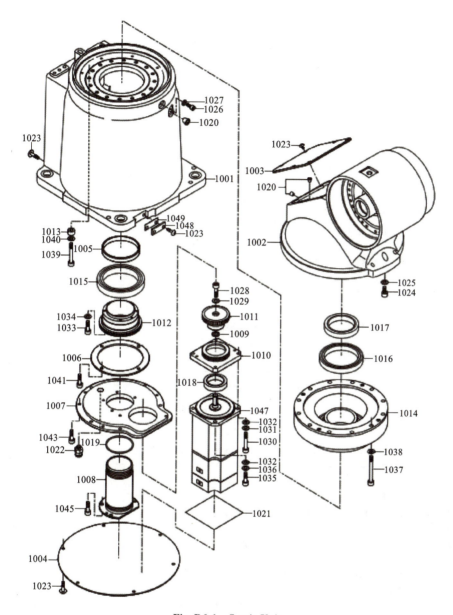

Fig. R6-1 S-axis Unit

Table R6-1 S-axis Unit

No.	Name	Pcs	No.	Name	Pcs
1001	Base	1	1024	Socket screw	1
1002	S head	1	1025	Spring washer	1

续表

No.	Name	Pcs	No.	Name	Pcs
1003	Cover	1	1026	M6X8 Socket screw	1
1004	Cover	1	1027	2H-6 Spring washer	1
1005	Collar	1	1028	M5X16 Socket screw	1
1006	B holder	1	1029	2H-5 Spring washer	1
1007	M base	1	1030	M4X25 Socket screw	2
1008	Shaft	1	1031	2H-4 Spring washer	2
1009	Packing	1	1032	M4 Washer	4
1010	M base	1	1033	M3X30 Socket screw	6
1011	Gear	1	1034	2H-3 Spring washer	6
1012	Gear	1	1035	M4X14 Socket screw	2
1013	Washer	16	1036	2H-4 Spring washer	2
1014	Speed reducer	1	1037	M5X40 Socket screw	12
1015	Bearing	1	1038	2H-5 Spring washer	12
1016	Bearing	1	1039	M5X30 Socket screw	16
1017	Washer	1	1040	2H-5 Spring washer	16
1018	Oil seal	1	1041	M4X10 GT-SA bolt	4
1019	Washer	1	1043	M4X16 GT-SA bolt	7
1020	Plug	3	1045	M4X12 GT-SA bolt	4
1021	Sheet	1	1047	SGMAV-04ANA-YR11 Motor	1
1022	Union	1	1048	HW0414483-1 Cover	1
1023	Button bolt	22	1049	HW0414484-1 Packing	1

2. Part Names and Working Axes

Part names and working axes are shown in Fig. R6-2.

3. Manipulator Base Dimensions

Manipulator base dimensions are shown in Fig. R6-3.

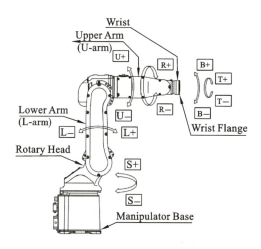

Fig. R6-2 Part Names and Working Axes

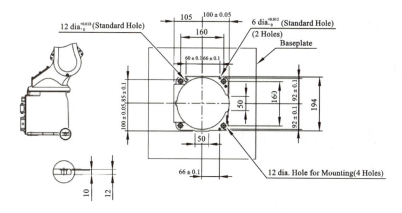

Fig. R6-3 Manipulator Base Dimensions

New Words and Phrases

washer [ˈwɒʃə(r)] n. (螺母等的)垫圈,垫片,衬垫,洗衣机
screw [skruː] n. 螺钉
gear [ɡɪə(r)] n. 齿轮,传动装置
motor [ˈməʊtə(r)] n. 发动机,引擎
wrist [rɪst] n. 手腕,腕关节
flange [flændʒ] n. 凸缘
mount [maʊnt] v. 安装,安置,裱
dimension [dɪˈmenʃn] n. 尺寸

Reading Material 7 Emergency Safety Information

7.1 Stop the system

1. Overview

Press any of the emergency stop buttons immediately if:

(1) There is personnel in the robot manipulator area, while the manipulator is working.

(2) The manipulator causes harm to personnel or mechanical equipment.

The FlexPendant emergency stop button is shown in Fig. R7-1.

Fig. R7-1 Emergency Stop Button

2. The Controller Emergency Stop Button

The emergency stop button on the controller is located on the front of the cabinet, as shown in Fig. R7-2, Fig. R7-3, and Fig. R7-4. However, this can differ depending on your plant design.

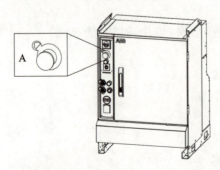

Fig. R7-2 Emergency Stop Button, Single Cabinet Controller

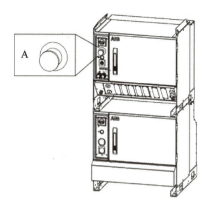

Fig. R7-3　Emergency Stop Button, Dual Cabinet Controller

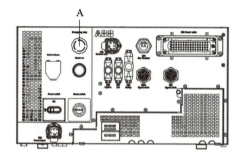

Fig. R7-4　Emergency Stop Button, IRC5 Compact

3. Other Emergency Stop Devices

The plant designer may have placed additional emergency stop devices in convenient places.

Please consult your plant or cell documentation to find out where these are placed.

7.2　Fire Extinguishing

Note: use a carbon dioxide (CO_2) extinguisher in the event of a fire in the manipulator system (manipulator or controller)!

dual cabinet controller
双机柜控制器

carbon dioxide
(CO_2)
二氧化碳
manipulator system
机械手系统

7.3　Shut Down All Power to the Controller

1. Overview

The controller has one main power switch on each module. To make sure that no power is connected to the controller, all modules' main switches must be turned off.

main power switch
主电源开关

2. Note

Your plant or cell may have additional equipment that may also need to be disconnected from the power. Please consult your plant or cell documentation to find out where these power switches are placed.

Shut down power to the controller, as shown in Table R7-1.

Table R7-1 Shut Down Power to the Controller

Step	Action	Info
1	Turn off the main power switch on the control module	If your system uses a single cabinet controller, a modular controller, or an IRC5 compact controller, then only step 1 is necessary
2	Turn off the main power switch on any connected drive module and other modules, such as spot welding cabinets, etc.	Please refer to the illustration

control module
控制模块
drive module
驱动模块
spotwelding cabinet
点焊机柜
additional cabinet
附加机柜

7.3.1 Main Power Switch, Single Cabinet Controller

The main power switch is located on the front of the controller cabinet for the single cabinet controller and IRC5 compact controller, as shown in Fig. R7-5. The design of a modular controller depends on the installation.

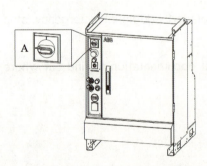

Fig. R7-5 Main Power Switch, Single Cabinet Controller

7.3.2 Main Power Switch, Controller with Additional Cabinets

Note that each connected drive module, or other connected modules such as a spot welding cabinet, has its own main power switch. On other types of cabinets, the power switch is often placed top left on the front of the cabinet, as shown in Fig. R7-6.

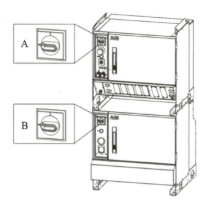

Fig. R7-6 Main Power Switch, Control Module (Dual Cabinet Controller)
A—main power switch, control module (dual cabinet controller);
B—main power switch, drive module (dual cabinet controller)

7.4 Releasing a Person Trapped by the Robot Arm

1. Overview

If a person has been trapped by the robot arm, you must make sure that any attempt to release the person does not further increase the risk of injury.

Releasing the robot holding brakes will make it possible to move the robot manually, but only small robots are light enough to be moved by human force. Moving larger robots may require using an overhead crane or similar. Make sure you have the right equipment at hand before releasing the brakes.

holding brake
保持制动器
overhead crane
桥式起重机

2. Warning

Before releasing the brakes, make sure that the weight of the arms does not increase the pressure on the trapped person, further increasing any injury!

7.5 How to Release a Trapped Person

Table R7-2 details how to release a person trapped by the robot arm.

Table R7-2 How to Release a Person Trapped by the Robot Arm

Step	Action
1	Press any of the emergency stop buttons
2	Make sure that the trapped person will not be more injured by the intended release action

续表

Step	Action
3	Move the robot so that the trapped person is released
4	Help the trapped person and make sure he or she gets medical attention
5	Make sure the robot cell is cleared so that no one else runs the risk of being injured

New Words and Phrases

emergency [ɪˈmɜːdʒənsi] n. 突发事件，紧急情况

button [ˈbʌt(ə)n] n. 按钮，纽扣，扣子

consult [kənˈsʌlt] v. 咨询，请教，商量，商讨

cell [sel] n. 小隔间，细胞

documentation [ˌdɒkjumenˈteɪʃn] n. 文件，凭证

extinguishing [ɪkˈstɪŋgwɪʃɪŋ] n. 熄灭 v. 熄灭，(使)消亡，破灭

switch [swɪtʃ] v. 打开，改变，转变

installation [ˌɪnstəˈleɪʃn] n. 安装，设置，装置，设备

release [rɪˈliːs] v. 释放，放走，松开

trap [træp] v. 使受限制，把……困在

brake [breɪk] n. 刹车，车闸

manually [ˈmænjuəli] adv. 手动地，用手

injured [ˈɪndʒəd] adj. 受伤的，有伤的，委屈的

FlexPendant 示教器

emergency stop button 紧急停止按钮

single cabinet controller 单机柜控制器

dual cabinet controller 双机柜控制器

carbon dioxide (CO_2) 二氧化碳

manipulator system 机械手系统

main power switch 主电源开关

control module 控制模块

drive module 驱动模块

spotwelding cabinet 点焊机柜

additional cabinet 附加机柜

overhead crane 桥式起重机

holding brake 保持制动器

Reading Material 8　　The Robot Palletizer

With the progress of science, technology and the acceleration of the modernization process, people have higher and higher requirements for handling speed, and the traditional manual stacking can only be used in the occasions of material portability, large size and shape change, small throughput, which has been far from meeting the needs of the industry, robot stacking came into being. Palletizing robot (as shown in Fig. R8-1) extends and expands the functions of human hands, feet, and brain. It can replace porters to work in dangerous, toxic, low temperature, high heat, and other harsh environments. Help people to complete heavy, monotonous, repetitive work, improve labor productivity, and ensure the quality of products.

Margin Note
robot palletizer
机器人码垛工
modernization process
现代化进程
palletizing robot
码垛机器人
labor productivity
劳动生产率

Fig. R8-1　The Palletizing Robot

The stacking robot designed and manufactured by combining mechanical devices and computer programs can work flexibly, accurately, quickly, stably, and efficiently, greatly reducing the overall labor cost and increasing commercial profit. Equipped with a robot packing and robot stacking system, it can accomplish complex tasks that previously required a lot of manpower, such as packing, stacking, unstacking, and order picking. Manufacturers can provide reliable and efficient personalized solutions according to customers' needs. The robot stacking system is very flexible, and can handle single, double, or multiple packages of various packaging sizes, shapes, or weights, including rigid or flexible packaging in the form of wooden cases, cartons, bags, handbags, drums, strapping, freight boards, or pallets. The

commercial profit
商业利润

weight range of stowage cargo is also very wide, the payload range is from 0.5 kg to 2 300 kg, suitable for any application or industry. Such as robot palletizers have been developed to design and manufacture specialized integrated equipment for the food, paint, coatings, chemical, petroleum, personal care, and beverage industries.

freight board
运货板
stowage cargo
码垛货物
payload range
载荷范围

The stacking robot is divided into vertical or horizontal loading systems to adapt to the stacking of various forms of product packaging. The space occupied by robot installation is flexible and compact. For example, the stacking robot of the suitcase system adopts a high-speed robot, with small and compact installation space. The system uses high-level programming language and an advanced controller, which is characterized by short programming time, fast processing speed, and convenient language programming.

With the increasingly high requirements for palletizing robots in terms of speed, accuracy, and product characteristics, technical innovations stand out. Engineers use SOLIDWORKS software to monitor the design process and improve product drawings through 3D visualization technology.

technical innovation
技术创新
visualization technology
可视化技术

As one of the important parts of the palletizing robot, the palletizing robot has high reliability, simple and novel structure, small quality, and other parameters, which is of great significance to the overall performance of the palletizing robot. Commonly used palletizing machine hand grasp mainly includes the following kinds:

(1) Clip-and-grab mechanical gripper (as shown in Fig. R8-2): the clip-and-grab mechanical gripper is mainly used for palletizing bags at high speed, this kind of manipulator is mainly used for bag stacking, such as flour, feed, cement, fertilizer, and so on.

clip-and-grab mechanical gripper
夹抓式机械抓手

Fig. R8-2　Clip-and-grab Mechanical Gripper

(2) Splint mechanical gripper (as shown in Fig. R8-3): the splint mechanical gripper is mainly suitable for stacking boxes. This kind of gripper is mainly used for stacking boxes or regular boxes of packaged

splint mechanical gripper
夹板式机械抓手

goods, which can be used in various industries. One or more boxes can be loaded at a time.

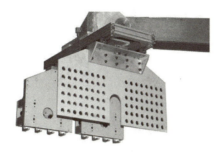

Fig. R8-3 Splint Mechanical Gripper

(3) Vacuum suction mechanical gripper (as shown in Fig. R8-4): the vacuum suction mechanical gripper is mainly suitable for suction of stacking objects, this kind of gripper is mainly used for stacking objects suitable for suction, such as film-covered packaging boxes, cans of beer, plastic boxes, cartons, and so on.

(4) Combined mechanical gripper (as shown in Fig. R8-5): the combined mechanical gripper is suitable for the cooperation of several stations, and is a flexible combination of the first three kinds of grippers, at the same time to meet the needs of multiple stations.

vacuum suction mechanical gripper 真空吸取式机械抓手
film-covered packaging box 覆膜包装盒
combined mechanical gripper 组合式机械抓手

Fig. R8-4 Vacuum Suction Mechanical Gripper

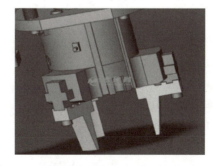

Fig. R8-5 Combined Mechanical Gripper

 New Words and Phrases

modernization [ˌmɒdənaɪˈzeɪʃn] n. 现代化
stacking [ˈstækɪŋ] n. 堆垛 v. 堆叠，堆积

throughput [ˈθruːpʊt] n. 吞吐量,生产量,接待人数
porter [ˈpɔːtə(r)] n. 搬运工人,脚夫
monotonous [məˈnɒtənəs] adj. 单调乏味的,毫无变化的
flexibly [ˈfleksəbli] adv. 灵活地,柔软地,有弹性地
strapping [ˈstræpɪŋ] n. 捆扎,皮带
payload [ˈpeɪləʊd] n. 净负荷,负载
coating [ˈkəʊtɪŋ] n. 涂料,涂层
petroleum [pəˈtrəʊliəm] n. 石油,原油
beverage [ˈbevərɪdʒ] n. 饮料
clip [klɪp] n. 夹子,别针,片段
gripper [ˈɡrɪpə(r)] n. 夹具,钳子
splint [splɪnt] n. 夹板,薄木条
cement [sɪˈment] n. 水泥
fertilizer [ˈfɜːtəlaɪzə(r)] n. 化肥,肥料
vacuum [ˈvækjuːm] adj. 真空的 n. 真空容器
suction [ˈsʌkʃ(ə)n] n. 吸盘,吸力,抽吸
plastic [ˈplæstɪk] adj. 塑料的,人造的,不自然的
carton [ˈkɑːtn] n. 硬纸盒,塑料盒
productivity [ˌprɒdʌkˈtɪvəti] n. 生产率,生产力
robot palletizer 机器人码垛工
modernization process 现代化进程
palletizing robot 码垛机器人
labor productivity 劳动生产率
freight board 运货板
stowage cargo 码垛货物
payload range 载荷范围
visualization technology 可视化技术
clip-and-grab mechanical gripper 夹抓式机械抓手
splint mechanical gripper 夹板式机械抓手
vacuum suction mechanical gripper 真空吸取式机械抓手
film-covered packaging box 覆膜包装盒
combined mechanical gripper 组合式机械抓手

Unit Ⅲ　Manufacturing Technology Informatization

Introduction to the Unit 单元导言

> 信息化是智能制造的必经之路。智能制造是基于新一代信息技术的先进制造过程、系统与模式的总称。本单元将重点讨论有关智能制造信息化方面的内容，主要是工业无线连接，以及数字孪生技术和增强现实技术在制造业中的应用；还介绍了生产一线常用的自动导引车、用于B型控制器的示教器的工作原理等内容。通过本单元的学习，学生可以理解信息技术与制造技术的关系，并通过操作手册掌握机器人示教器使用方法；还可以进一步理解制造业向数字化、智能化发展的意义，初步掌握信息技术和制造技术的英语应用。

Lesson 11　Wireless Connectivity for Industries

Industry 4.0 refers to the shift in the way products are produced and delivered to industrial automation and flexible factories.

To stay competitive, factories and warehouses must leverage the industrial internet of things (IIoT), as shown in Fig. 11-1, and digitalization to become much more agile and efficient.

Margin Note
industrial automation
工业自动化
flexible factory
柔性工厂

Fig. 11-1　The Industrial Internet of Things

While industries have automated many processes, secure wireless connectivity empowers factory automation, making industrial automation possible on a much larger scale.

Huge gains await industries that cut the cord and go wireless. Wireless cellular connectivity supports the business outcomes that the industry expects from Industry 4.0. For instance, manufacturing enables flexible production by allowing smart factories to rapidly change over production lines to shorten lead times. Moving from wired to wireless cellular connectivity brings greater flexibility to Industry 4.0 operations.

wireless cellular connectivity
无线蜂窝连接
lead time
订货交付时间

Built on the foundation of smart, secure, wireless connectivity, there are opportunities to extend machine life through predictive maintenance (as shown in Fig. 11-2), support rapid material handling, monitor every detail of the shop floor, and leverage collaborative robots simultaneously with mobile communication. This will help factories realize their goal of becoming fully automated factories. Industry 4.0 will help make smart machines smarter, factories more efficient, processes less wasteful, production lines more flexible, and productivity higher.

Fig. 11-2　Predictive Maintenance

For any industrial facility requiring operational optimization through automation, control, and insights from data analysis, a private cellular network utilizes radio coverage to become independent of[①] public mobile networks, ensuring it meets cellular capacity that encompasses high device density and predictable latency and allows for[②] the smooth transition to 5G.

private cellular network
专用蜂窝网络

The main benefits of private networks are:

(1) Guaranteed coverage is the most apparent benefit. As shown in Fig. 11-3, it is assured in the enterprise's operations area, both indoors and outdoors even in remote locations as required, through a dedicated

private network
专用网络

spectrum.

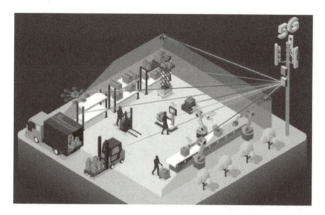

Fig. 11-3 Guaranteed Coverage of Private 5G Network

(2) Security and encryption are vital and of the strongest standards. Since private networks are based on③ cellular technology, they can offer high levels of security as 3GPP standards are closely adhered to④ across vendors, and the private nature of this solution ensures that all data stays on-premises.

(3) Ensured capacity is an essential feature, as a private network removes any contention with other network users, making it possible to guarantee network performance, such as uplink and downlink bit rates and latency.

(4) Retained control due to⑤ private networks enables enterprises to determine and control how resources are utilized, and how traffic is prioritized. The security of the radio access network can be controlled to ensure that sensitive information remains on-premises.

(5) Critical reliability is assured as private networks are based on LTE/5G technology, which offers performance and enables applications that cannot be accommodated by Wi-Fi, such as mission-critical communication services and ultra-high definition video surveillance.

(6) Predictable and ensured low latency is another important feature. It is a requirement for many internet of things(IoT) applications that rely on⑥ time-bound communications, where delays can result in a catastrophic failure, such as for critical control of remote devices like heavy machinery.

(7) High data speeds for communication are offered. This is required for video and high-resolution images in many industry sectors.

uplink
上行链路
downlink
下行链路

radio
access network
无线接入网

ultra-high definition
超高清
low latency
低时延

Tips

1. 3GPP:3rd Generation Partnership Project 第三代合作伙伴计划

第三代合作伙伴计划成立于 1998 年 12 月，最初是为第三代移动通信系统制定全球适用的技术规范和技术报告。随后，工作范围增加了对 UTRA（universal terrestrial radio access,通用地面无线接入）长期演进系统的研究和标准制定。其组织伙伴包括 ETSI（European Telecommunications Standards Institute,欧洲电信标准化协会）、ARIB（Association of Radio Industries and Businesses,日本无线工业及商贸联合会）、CCSA（China Communications Standards Association,中国通信标准化协会）等七个标准化组织。

2. bit rates 比特率

比特率是指单位时间内传送的比特（bit）数。比特率单位为 bps（bit per second），也可表示为 bit/s,比特率越高，单位时间内传送的数据量（位数）越大。计算机中的信息都用二进制的 0 和 1 来表示，其中每一个 0 或 1 被称作一个位，用小写 b 表示，即 bit（位）。大写 B 表示 byte，即字节，1 个字节＝8 个位，即 1 B＝8 b。一般使用千字节（KB）来表示文件的大小。

New Words and Phrases

flexible factory 柔性工厂
warehouse [ˈweəhaʊs] n. 仓库,货栈,货仓
agile [ˈædʒaɪl] adj. 灵活的,敏捷的,机敏的,机灵的
cellular [ˈseljələ(r)] adj. （无线电话）蜂窝状的,细胞的,网状的
private cellular network 专用蜂窝网络
wireless cellular connectivity 无线蜂窝连接
lead time 订货交付时间
simultaneously [ˌsɪmlˈteɪniəsli] adv. 同时地,同步地
optimization [ˌɒptɪmaɪˈzeɪʃn] n. 最优化,充分利用
utilize [ˈjuːtəlaɪz] v. 利用,使用,运用,应用
encompass [ɪnˈkʌmpəs] v. 包含,涉及（大量事物）,包围,围绕
private network 专用网络
dedicated [ˈdedɪkeɪtɪd] adj. 专用的,专门用途的,献身的,专心的

spectrum ['spektrəm] n. 频谱,声谱,波谱,光谱,谱,范围,幅度
encryption [ɪn'krɪpʃn] n. 加密,加密技术
vendor ['vendə(r)] n. 供应商,销售公司,摊贩,(房屋等的)卖主
on-premises 本地部署的(系统)
uplink ['ʌplɪŋk] n. 上行链路
downlink ['daʊnlɪŋk] n. 下行链路
contention [kən'tenʃn] n. 争用,争吵,争论,看法,观点
prioritize [praɪ'ɒrətaɪz] v. 划分优先顺序,优先处理,按重要性排列
radio access network 无线接入网
critical ['krɪtɪkl] adj. 极其重要的,关键的,批判的
mission-critical 关键的,至关重要的
definition [ˌdefɪ'nɪʃn] n. 清晰,清晰度,定义,释义,榜样,典范
ultra-high definition 超高清
surveillance [sɜː'veɪləns] n. 监控,监视
latency ['leɪtənsi] n. 时延
low latency 低时延
catastrophic [ˌkætə'strɒfɪk] adj. 灾难性的

Notes

1. independent of 独立于……之外,不受……支配,与……无关,不依赖……

例句:For any industrial facility requiring operational optimization through automation, control, and insights from data analysis, a private cellular network utilizes radio coverage to become independent of public mobile networks, ensuring it meets cellular capacity that encompasses high device density and predictable latency and allows for the smooth transition to 5G.

对于任何需要通过自动化、控制和数据分析进行操作最优化的工业设施,专用蜂窝网络可利用无线电覆盖使其从公共移动网络中独立出来,确保它能达到包括高设备密度和可预测时延的蜂窝容量,且能考虑到向 5G 的平稳过渡。

例句:We need a central bank that is independent of the government.
我们需要一个不受政府控制的独立的中央银行。

2. allow for 考虑到,把……计算在内

例句:For any industrial facility requiring operational optimization through automation, control, and insights from data analysis, a private cellular network utilizes radio coverage to become independent of public mobile networks, ensuring it meets cellular capacity that encompasses high device density and predictable latency and allows for the smooth transition to 5G.

对于任何需要通过自动化、控制和数据分析进行操作最优化的工业设施,专用蜂窝网络可利用无线电覆盖使其从公共移动网络中独立出来,确保它能达到包括高设备密度和可预测时延的蜂窝容量,且能考虑到向 5G 的平稳过渡。

例句:All these factors must be allowed for.

这些因素都必须考虑到。

3. be based on 以……为基础(或根据)

例句:Since private networks are based on cellular technology, they can offer high levels of security as 3GPP standards are closely adhered to across vendors.

由于专用网络的构建基于蜂窝技术,且各个供应商都严格遵守 3GPP 标准,因此专用网络能提供高级别的安全防护。

例句:The movie is based on a real-life incident.

这部电影以真实事件为蓝本。

4. adhere to 遵循,坚持,遵守(法律、规章、指示、信念等),附着,黏附

例句:Since private networks are based on cellular technology, they can offer high levels of security as 3GPP standards are closely adhered to across vendors.

由于专用网络的构建基于蜂窝技术,且各个供应商都严格遵守 3GPP 标准,因此专用网络能提供高级别的安全防护。

例句:Small particles adhere to the seed.

小颗粒牢牢地附着在种子上。

5. due to 由于,因为

例句:Retained control due to private networks enables enterprises to determine and control how resources are utilized, and how traffic is prioritized.

由于专用网络保留了控制权,企业得以确定并控制资源的使用以及流量的排序。

例句:The train stopped due to a mechanical problem.

火车因为机械故障停了下来。

6. rely on 依赖,依靠,信任,信赖

例句:It is a requirement for many internet of things(IoT) applications that rely on time-bound communications, where delays can result in a catastrophic failure, such as for critical control of remote devices like heavy machinery.

它是许多依赖于限时通信的物联网应用运行的必备条件,在这些应用中,延迟可能导致灾难性的故障,例如对重型机械等远程设备的关键控制。

例句:You should rely on your judgment.

你应该相信自己的判断。

Unit Ⅲ   Manufacturing Technology Informatization | 97

 *Exercises*

Ⅰ. Write True or False beside the following statements about the text.

1. _____ Moving from wired to wireless cellular connectivity brings greater flexibility to Industry 4.0 operations.

2. _____ Industry 4.0 will help make smart machines smarter while processes will be more wasteful.

3. _____ Guaranteed coverage is the most apparent benefit of private networks.

4. _____ Few vendors adhere 3GPP standards.

5. _____ High data speeds for communication are offered, which are required for video and high-resolution images in many industry sectors.

Ⅱ. Answer the following questions in English according to the text.

1. What does Industry 4.0 refer to? How can factories and warehouses stay competitive?

2. Why will industries that cut the cord and go wireless have huge gains?

3. What are the main benefits of private networks?

Ⅲ. Read the text again and fill in the blanks in the following sentences orally.

1. Industry 4.0 refers to the shift in the way products are produced and delivered to _____ _____ and _____ _____.

2. Moving from wired to wireless _____ connectivity brings greater flexibility to Industry 4.0 operations.

3. For any industrial facility requiring operational optimization through automation, control, and insights from data analysis, a _____ _____ network utilizes radio coverage to become independent of _____ _____ networks, ensuring it meets cellular capacity that encompasses high device density and predictable latency and allows for the smooth transition to 5G.

4. Security and _____ are vital and of the strongest standards. Since private networks are based on cellular technology, they can offer high levels of _____ as 3GPP standards are closely adhered to across vendors, and the private nature of this solution ensures that all data stays _____.

5. Predictable and ensured low latency is another important feature. It is a requirement for many _____ _____ _____ applications that rely on _____ communications, where delays can result in a _____ failure, such as for critical control of remote devices like heavy machinery.

Ⅳ. Translation.

Huge gains await industries that cut the cord and go wireless. Wireless cellular connectivity supports the business outcomes that the industry expects from Industry 4.0. For instance, manufacturing enables flexible production by allowing smart factories to rapidly change over production lines to shorten lead times. Moving from wired to wireless cellular connectivity brings greater flexibility to Industry 4.0 operations.

Lesson 12 Application of Digital Twin Technology in Manufacturing Industry

Digital twin, also known as digital mapping or digital mirroring, refers to① the simulation of physical objects, processes, or systems in the information platform. It is similar to② the twins of physical systems in the information platform. The essence of the digital twin is information modeling, which aims to build a completely consistent digital model in the digital virtual world for the physical objects in the real world, as shown in Fig. 12-1.

Margin Note
digital twin
数字孪生
digital mapping
数字映射
digital mirroring
数字镜像
information platform
信息平台
information modeling
信息建模

Fig. 12-1　Physical Objects and Digital Models

Digital twin in the workplace is often considered part of robotic process automation (RPA). An example of the digital twin is the use of 3D modeling to create digital companions for physical objects. It can be used to view the status of the actual physical object, which provides a way to project physical objects into the digital world. For example, when sensors collect data from a connected device, the sensor data can be used to update a "digital twin" copy of the device's state in real-time. The term "device shadow" is also used for the concept of the digital twin. The digital twin is meant to be③ an up-to-date and accurate

digital companion
数字伙伴

device shadow
设备影子

copy of the physical object's properties and states, including shape, position, gesture, status, and motion. It can also be used for monitoring, diagnostics, and prognostics to optimize asset performance and utilization. In this field, sensory data can be combined with④ historical data, human expertise, and simulation learning to improve the outcome of prognostics.

The characteristics of digital twin technology are:

(1) Virtual real mapping. Digital twin technology requires the construction of digital representations of physical objects in the digital space. Physical objects in the real world and the twin in the digital space can realize bidirectional mapping, data connection, and state interaction.

(2) Real-time synchronization. Based on the acquisition of real-time sensing and other multivariate data, the twin body can comprehensively, accurately, and dynamically reflect the state changes of physical objects, including appearance, performance, position, anomaly, etc.

(3) Symbiotic evolution. In an ideal state, the mapping and synchronization state realized by the digital twin should cover the whole life cycle of the twin object from design, production, and operation to scrap, and the twin object should evolve and be updated with its life cycle process.

(4) Closed-loop optimization. The ultimate purpose of establishing the twin body is to form optimization instructions or strategies for the physical world based on analysis and simulation by describing the internal mechanism of the physical object, analyzing the laws and insights into the trend, so as to⑤ realize the closed-loop decision-making optimization function of the physical object.

Digital twin technology can be applied at all stages of intelligent manufacturing, including design, production, and operation and maintenance.

(1) Design phase. Through collaborative design, people create a design model, a digital twin, for the products to complete product design. For manufacturability, the virtual environment can be used for simulation verification, and then enter the production stage.

(2) Production phase. The digital twin of a physical factory operates in the virtual space of the cloud platform. Meanwhile, physical devices and the internet of things composed of sensors obtain forecast data through

sensory data
感官数据

virtual real mapping
虚实映射

real-time synchronization
实时同步
real-time sensing
实时传感

symbiotic evolution
共生演进

life cycle
生命周期
closed-loop optimization
闭环优化

simulation verification
仿真验证

model interaction
模型交互

a low latency network using model interaction and complete the optimal control of the machining process within the control cycle.

(3) Operation and maintenance phase. The digital twin can not only realize the real-time monitoring of product use status but also be used in operation, training, and guidance. With the help of virtual reality fusion technology, it can also provide more realistic effects.

In a word, the technology of digital twin is a new technology born on the basis of⑥ traditional simulation technology under the background of booming development of cloud computing, big data, 3D modeling, artificial intelligence, industrial internet of things, and deep learning. The future of manufacturing will be driven by the following four aspects: modularity, autonomy, connectivity, and digital twin.

virtual reality fusion technology
虚拟现实融合技术

cloud computing
云计算

artificial intelligence
人工智能

deep learning
深度学习

Tips

1. RPA: Robotic Process Automation 机器人流程自动化

机器人流程自动化是指用软件自动化方式实现各个行业中本来由人工操作计算机完成的业务。在企业的业务流程中，通常有纸质文件录入、证件票据验证、从电子邮件和文档中提取数据、跨系统数据迁移、企业信息技术（IT）应用自动操作等。软件机器人能准确、快速地完成工作、减少人工错误、确保零失误、提高效率、大幅降低运营成本，实现了企业内部跨系统及工作流程的自动化工作。

2. device shadow 设备影子

设备影子是一个 JSON（JavaScript object notation, JavaScript 对象简谱）文档，用于存储设备上报状态、应用程序期望状态信息。每个设备有且只有一个设备影子，设备可以通过 MQTT（message queuing telemetry transport, 消息队列遥测传输）协议获取和设置设备影子同步。该同步可以是影子同步给设备，也可以是设备同步给影子。物联网平台提供设备影子功能，用于缓存设备状态。

3. deep learning 深度学习

深度学习是机器学习领域中一个新的研究方向。它是学习样本数据的内在规律和表示层次，这些在学习过程中获得的信息对文字、图像和声音等数据的解释有很大的帮助。它的最终目标是让机器能够像人一样具有分析学习能力，能够识别文字、图像和声音等数据。深度学习应用在搜索技术、数据挖掘、机器学习、机器翻译、自然语言处理、推荐的模式识别等领域，使人工智能相关技术取得了很大进步。

New Words and Phrases

mapping [ˈmæpɪŋ] n. (数学、语言学)映射,映现
digital mapping 数字映射
mirroring [ˈmɪrərɪŋ] n. 反射,镜面反射
digital mirroring 数字镜像
information platform 信息平台
information modeling 信息建模
digital companion 数字伙伴
shadow [ˈʃædəʊ] n. 影子,阴影 v. 被……阴影笼罩
device shadow 设备影子
sensory data 感官数据
virtual real mapping 虚实映射
synchronization [ˌsɪŋkrənaɪˈzeɪʃn] n. 同步
real-time synchronization 实时同步
real-time sensing 实时传感
multivariate [ˌmʌltɪˈveərɪɪt] adj. 多元的,多变量的
symbiotic [ˌsɪmbaɪˈɒtɪk] adj. 共生的,互利的
symbiotic evolution 共生演进
life cycle 生命周期
scrap return 回炉料
closed-loop optimization 闭环优化
manufacturability [ˌmænjʊfæktʃərəˈbɪlɪtɪ] n. 可制造性,工艺性,可生产性
verification [ˌverɪfɪˈkeɪʃ(ə)n] n. 验证,证明
simulation verification 仿真验证
model interaction 模型交互
fusion [ˈfjuːʒn] n. 融合,结合
virtual reality fusion technology 虚拟现实融合技术
cloud computing 云计算
modularity [ˌmɒdjʊˈlærɪtɪ] n. 模块化,模块性
autonomy [ɔːˈtɒnəmɪ] n. 自主化,自主,自主权

Notes

1. **refer to** 指的是……,涉及,查阅,参考,提到,谈及

例句: Digital twin, also known as digital mapping or digital mirroring, refers to the simulation of physical objects, processes, or systems in the information platform.

数字孪生又被称作数字映射、数字镜像,指的是在信息平台上对物理对象、过程或者系统的模拟。

例句:The item "arts" usually refers to humanities and social sciences.
"arts"一词通常指人文和社会科学。

2. similar to 类似于,与……相像

例句:It is similar to the twins of physical systems in the information platform.
它类似于信息平台上物理系统的孪生。

例句:The basic design of the car is very similar to that of earlier models.
这种汽车的基本设计与早期的样式非常相似。

3. be meant to be 被普遍认为是……

例句:The digital twin is meant to be an up-to-date and accurate copy of the physical object's properties and states, including shape, position, gesture, status, and motion.
数字孪生被普遍视作物理对象属性和状态的最新精准副本,其中包括形状、位置、姿势、状态和运动。

例句:This restaurant is meant to be excellent.
都说这家饭店很棒。

4. be combined with 与……结合/联合

例句:In this field, sensory data can be combined with historical data, human expertise, and simulation learning to improve the outcome of prognostics.
在这一领域,感官数据可以与历史数据、人类专业知识和模拟学习相结合,以改善预测结果。

例句:It can be combined with other herbs.
它可以与其他草药结合使用。

5. so as to 为了(做某事),以便(做某事)

例句:The ultimate purpose of establishing the twin body is to form optimization instructions or strategies for the physical world based on analysis and simulation by describing the internal mechanism of the physical object, analyzing the laws and insights into the trend, so as to realize the closed-loop decision-making optimization function of the physical object.
建立孪生体的最终目的是通过描述物理对象内在机理、分析规律、洞察趋势,基于分析与仿真对物理世界形成优化指令或策略,以实现对物理对象决策优化功能的闭环。

例句:We went early so as to get good seats.
为了占到好座位,我们早早就去了。

6. on the basis of 在……基础上,依据……,根据……,出于……原因

例句:The technology of digital twin is a new technology born on the basis of traditional simulation technology under the background of booming development of cloud computing, big data, 3D modeling, artificial intelligence, industrial internet of things, and deep learning.

数字孪生技术是在云计算、大数据、三维建模、人工智能、工业物联网和深度学习蓬勃发展的背景下,<u>在传统仿真技术的基础上</u>诞生的一项新技术。

例句:The denunciation was made <u>on the basis of</u> second-hand information.

这些指责是<u>依据</u>二手信息提出的。

 Exercises

Ⅰ. **Write True or False beside the following statements about the text.**

1. _____ Digital twin is also known as digital mapping or digital mirroring.

2. _____ Digital twin in the workplace is often considered part of robotic process automation (RPA).

3. _____ The term "device shadow" has nothing to do with the concept of the digital twin.

4. _____ Digital twin also can be used for monitoring, diagnostics, and prognostics to optimize asset performance and utilization.

5. _____ Digital twin technology can be applied in the design phase, the production phase, and the operation and maintenance phase.

Ⅱ. **Answer the following questions in English according to the text.**

1. What does digital twin refer to?

2. What's the essence of the digital twin?

3. What are the characteristics of digital twin technology?

Ⅲ. **Read the text again and fill in the blanks in the following sentences orally.**

1. The digital twin is meant to be an _____ and _____ copy of the physical object's _____ and _____, including shape, position, gesture, status, and motion.

2. Digital twin also be used for _____, _____, and _____ to _____ asset performance and utilization. In this field, _____ _____ can be combined with historical data, human expertise, and simulation learning to improve the outcome of prognostics.

3. Based on the _____ of real-time sensing and other _____ data, the twin body can comprehensively, accurately, and dynamically reflect the state changes of physical objects, including _____, performance, position, _____, etc.

4. The digital twin can not only realize the real-time _____ of product use status but also be used in _____, _____ and _____. With the help of _____ _____ _____ technology, it can also provide more realistic effects.

5. The technology of digital twin is a new technology born on the basis of traditional

simulation technology under the background of booming development of _____ _____, _____ _____, 3D modeling, _____ _____, industrial internet of things, and _____ _____.

Ⅳ. Translation.

The ultimate purpose of establishing the twin body is to form optimization instructions or strategies for the physical world based on analysis and simulation by describing the internal mechanism of the physical object, analyzing the laws and insights into the trend, so as to realize the closed-loop decision-making optimization function of the physical object.

Lesson 13 Application of AR Technology in Manufacturing Industry

Augmented reality (AR) is a new area of research extending from virtual reality (VR) which is a combination of the virtual environment and the real environment. It is a technology that adds virtual objects generated by the computer to① the real world to construct the effect of the actual situation.

Augmented reality technology will overlay the position and angle of the real-time image of the camera with corresponding images, and integrate the information of the real world and the virtual world "seamlessly". The purpose is to set the virtual world on the screen and interact with the real world. Augmented reality includes multimedia, 3D modeling, real-time video display, multi-sensor fusion, real-time tracking, scene fusion, and other new technologies. AR has wide application prospect in the manufacturing industry.

1. Complex Assembly

Whether it's making a smartphone or a jet engine, hundreds or thousands of components need to be arranged and assembled quickly in a precise order. Each new product requires a new set of assembly instructions, and the workload is very large. As Fig. 13-1 shows, the modern manufacturing industry projects assembly information on② the display to guide the assembly process, which requires the use of augmented reality technology.

2. Equipment Maintenance

Augmented reality can also be used to aid in equipment

Margin Note
augmented reality
增强现实
virtual reality
虚拟现实

multi-sensor fusion
多传感器融合
real-time tracking
实时跟踪
scene fusion
场景融合
complex assembly
复杂装配

Fig. 13-1　Projecting Assembly Information on the Display to Guide the Assembly Process

maintenance. Based on 3D model augmented reality technology, the developer has developed a technical maintenance support AR system that presets various maintenance and inspection tasks. The augmented reality screen can help technicians to view machine status and detect errors. For example, as shown in Fig. 13-2, the technical personnel of the elevator manufacturer can view and identify the condition of the elevator before working with holographic technology, and can also carry out[3] remote technical access and expert consultation. AR equipment has been well applied in the industrial environment.

technical personnel
技术人员
holographic technology
全息技术

Fig. 13-2　Holographic Technology in Elevator Maintenance

3. Quality Assurance

Technicians have used augmented reality as a tool in the quality assurance process. At car factories, technicians take pictures of the parts of the vehicles being inspected, and then compare these images with standard images provided by the company through augmented reality technology, as shown in Fig. 13-3. Unqualified parts are highlighted by stacking, enabling quick and intuitive identification of vehicle problems and greatly reducing inspection time. Generally, there

quality assurance
质量保证

is a cage around the robot automation unit in the workshop, where a group of robots moves and work. Technicians can approach the automatic cell to check working conditions and get feedback from the augmented reality system by asking questions. Therefore, augmented reality technology has become an assistant tool for quality control.

quality control
品质控制

Fig. 13-3　Comparing the Parts with Standard Images

4. Expert Support

The skills gap in manufacturing could be filled with the help of augmented reality. Fig. 13-4 shows a new and effective training mode in which enterprises do not need to train technical personnel through off-duty training. Instead, they use the knowledge and technology of remote experts or expert systems in real-time through augmented reality technology to supplement the knowledge of employees on site④ dynamically and improve their skills.

Fig. 13-4　Expert Support

5. Automation

As manufacturing becomes increasingly automated, augmented

reality and the internet of things will replace human access to[5] information with[6] robots, making companies more competitive in the marketplace. With the development of technology, cameras on the production floor can also be directly connected to a cloud-based component database for real-time analysis, which is more conducive to[7] production automation.

production floor 生产车间
production automation 生产自动化

Tips

1. multi-sensor fusion 多传感器融合

多传感器融合一般指多传感器信息融合。它是利用计算机技术将来自多传感器或多源的信息和数据,在一定的准则下加以自动分析和综合,以完成所需要的决策和估计而进行的信息处理过程。其军事应用包括海洋监视系统和军事防御系统。在民事应用领域,多传感器融合主要用于智能处理以及工业控制。

2. holographic technology 全息技术

全息技术是利用干涉和衍射原理来记录并再现物体真实的三维图像的技术。所谓"全息",是指用投影的方法记录并且再现被拍物体发出的光的全部信息。光学全息术在立体电影、显微术、干涉度量学、投影光刻、军事侦察监视、水下探测、金属内部探测、信息存储、遥感、研究和记录物理状态变化极快的瞬时现象和瞬时过程(如爆炸和燃烧)等方面获得广泛应用。

New Words and Phrases

corresponding [ˌkɒrəˈspɒndɪŋ] *adj.* 相应的,符合的,相关的
multimedia [ˌmʌltiˈmiːdiə] *n.* 多媒体
multi-sensor fusion 多传感器融合
real-time tracking 实时跟踪
scene fusion 场景融合
holographic [ˌhɒləˈɡræfɪk] *adj.* 全息图的
holographic technology 全息技术
consultation [ˌkɒnslˈteɪʃn] *n.* 咨询,商讨,磋商,协商会
assurance [əˈʃʊərəns] *n.* 保证,确保
quality assurance 质量保证

stack [stæk] n.(使)放成整齐的一叠(一摞、一堆),许多,栈
quality control 品质控制
off-duty 非值勤的,歇班的
dynamically [daɪ'næmɪkli] adv.动态地
supplement ['sʌplɪmənt] v.增补,补充 n.增补物,增刊
marketplace ['mɑːkɪtpleɪs] n.市场,集市
production floor 生产车间
conducive [kən'djuːsɪv] adj.有助于(有利于)……的
production automation 生产自动化

 Notes

1. add sth. to sth. 将……添加进……,将……补充进……

例句:It is a technology that adds virtual objects generated by the computer to the real world to construct the effect of the actual situation.

它是一种将计算机生成的虚拟对象添加到现实世界中,以构建实际场景效果的技术。

例句:The suite will add a touch of class to your bedroom.

这套家具会给你的卧室增添一些典雅气息。

2. project sth. on sth. 将……投射(投影)在……上

例句:The modern manufacturing industry projects assembly information on the display to guide the assembly process, which requires the use of augmented reality technology.

现代制造业将装配信息投影到显示器上,指导装配工艺,这就需要用到增强现实技术。

例句:They were delighted to see their holiday slides projected on a screen.

他们很高兴看到自己的假日幻灯片在银幕上放映。

3. carry out 进行,实施,执行,完成(任务)

例句:For example, the technical personnel of the elevator manufacturer can view and identify the condition of the elevator before working with holographic technology, and can also carry out remote technical access and expert consultation.

举例来说,电梯制造厂家的技术人员可在工作开始前使用全息技术,查看并识别电梯的状况,还可以进行远程技术访问与专家咨询。

例句:Extensive tests have been carried out on the patient.

已对患者进行了全面检查。

4. on site 在现场

例句:Instead, they use the knowledge and technology of remote experts or expert systems in real-time through augmented reality technology to supplement the knowledge of employees on site dynamically and improve their skills.

企业无须通过下班培训训练技术人员,而是通过增强现实技术实时利用远程专家或专家系统的知识和技术,动态地补充现场员工的知识,提升其技能。

例句:Three CBS cameramen were on site to shoot and edit taped reports.

三名哥伦比亚广播公司的摄影记者在现场拍摄并编辑录像报道。

5. access to 获取,使用(查阅)……的机会(权利),去……的通路,进入,接近

例句:As manufacturing becomes increasingly automated, augmented reality and the internet of things will replace human access to information with robots, making companies more competitive in the marketplace.

制造业的自动化程度日益提高,增强现实技术和物联网技术将用机器人取代人类获取信息的途径,这将使企业在市场上更具竞争力。

例句:Students must have access to good resources.

学生必须有机会使用好的资源。

6. replace sth. with sth. 用……取代……,用……替换……

例句:As manufacturing becomes increasingly automated, augmented reality and the internet of things will replace human access to information with robots, making companies more competitive in the marketplace.

制造业的自动化程度日益增长,增强现实技术和物联网技术将用机器人取代人类获取信息的途径,这将使企业在市场上更具竞争力。

例句:It is not a good idea to miss meals and replace them with snacks.

不吃正餐,改吃点心,这不是什么好主意。

7. be conducive to 有利于……,有益于……

例句:With the development of technology, cameras on the production floor can also be directly connected to a cloud-based component database for real-time analysis, which is more conducive to production automation.

随着技术的发展,还可将生产车间的摄像头直接连接到云基础零部件数据库以实现实时分析,这将更加有利于实现生产自动化。

例句:Competition is not conducive to human happiness.

竞争不利于人的幸福。

 Exercises

Ⅰ. Write True or False beside the following statements about the text.

1. _____ Virtual reality (VR) is a new area of research extending from augmented reality (AR) which is a combination of the virtual environment and the real environment.

2. _____ AR equipment has not been applied in the industrial environment yet.

3. _____ Technicians can approach the automatic cell to check working conditions

and get feedback from the augmented reality system by asking questions.

4. _____ The skills gap in manufacturing could be filled with the help of augmented reality.

5. _____ Although manufacturing becomes increasingly automated, human access to information will never be replaced.

II. Answer the following questions in English according to the text.

1. What does AR refer to?
2. What technologies does AR include?
3. What are the applications of AR in manufacturing?
4. How can AR be used to aid in equipment maintenance?
5. How can AR be used in the quality assurance process?

III. Read the text again and fill in the blanks in the following sentences orally.

1. Augmented reality technology will _____ the position and angle of the _____ _____ of the camera with _____ _____, and _____ the information of the real world and the virtual world "seamlessly".

2. The modern manufacturing industry projects _____ _____ on the display to guide the _____ _____, which requires the use of augmented reality technology.

3. Based on 3D model augmented reality technology, the developer has developed a technical _____ support AR system that presets various maintenance and _____ tasks.

4. Technicians have used augmented reality as a tool in the _____ _____ process. At car factories, _____ take pictures of the parts of the vehicles being _____, and then compare these images with standard images provided by the company through augmented reality technology.

5. With the development of technology, cameras on the production floor can also be directly connected to a _____ _____ _____ for real-time analysis, which is more _____ to production automation.

IV. Translation.

As manufacturing becomes increasingly automated, augmented reality and the internet of things will replace human access to information with robots, making companies more competitive in the marketplace. With the development of technology, cameras on the production floor can also be directly connected to a cloud-based component database for real-time analysis, which is more conducive to production automation.

Lesson 14　Automated Guided Vehicles

　　An automated guided vehicle or automatic guided vehicle (AGV), also called an autonomous mobile robot (AMR), is a portable robot that follows along marked long lines or wires on the floor or uses radio waves, vision cameras, magnets, or lasers for navigation. AGVs are most often used in industrial applications to transport heavy materials around a large industrial building, such as a factory or a warehouse, as shown in Fig. 14-1.

Margin Note

Fig. 14-1　Automated Guided Vehicle

　　AGVs can tow objects behind them in trailers to which they can autonomously attach. The trailers can be used to move raw materials or finished products. AGVs are employed in nearly every industry, including pulp, paper, metals, newspaper, and general manufacturing.

　　Lower-cost versions of AGVs are often called AGCs (automated guided carts) and are usually guided by magnetic tape. AGCs are available in a variety of① models and can be used to move products on an assembly line, transport goods throughout a plant or a warehouse, and deliver loads, as shown in Fig. 14-2.

　　Over the years the technology has become more sophisticated and today automated vehicles are mainly laser-navigated, e.g. LGVs (laser guided vehicles). LGV is a type of AGV that is equipped with a laser navigation triangulation system, as shown in Fig. 14-3. In an automated process, LGVs are programmed to communicate with other robots to ensure products are moved smoothly through the warehouse, whether they are being stored for future use or sent directly to shipping areas.

magnetic tape
磁带

assembly line
流水线

laser guided vehicle
激光导引车

Fig. 14-2　Heavy-duty Automated Guided Cart

Today, the AGV plays an important role in② the design of new factories and warehouses, safely moving goods to the destination.

Fig. 14-3　Laser Guided Vehicle

　　In the wire-guided mode, a slot is cut in the floor and a wire is placed approximately 1 inch below the surface. As shown in Fig. 14-4, the slot is cut along the path that the AGV is to follow. The wire is used to transmit a radio signal. A sensor is installed on the bottom of the AGV close to the ground. It detects the relative position of the radio signal being transmitted from the wire. The information will be used to regulate the steering circuit, making the AGV follow the wire.

radio signal
无线电信号

　　In the tape-guided mode, AGVs (some known as AGCs) use tape as the guide path, as shown in Fig. 14-5. The tape can be magnetic or colored. The AGV is fitted with③ the appropriate guide sensor to follow the path of the tape. One major advantage of tape guidance over wire guidance is that it can be easily removed and relocated if the course needs to change. Colored tape is initially less expensive, but it lacks④ the advantage of being embedded in⑤ high-traffic areas where the tape

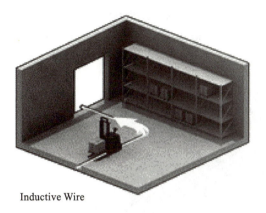

Inductive Wire

Fig. 14-4 Wire-guided Navigation

may become damaged or dirty. A flexible magnetic bar can be embedded in the floor like wire, but it works under the same provision as magnetic tape and so remains unpowered or passive. Another advantage of magnetic guide tape is its dual polarity. Small pieces of magnetic tape may be placed to change the state of the AGC based on polarity and the sequence of the tags.

dual polarity
双极性
laser target navigation
激光目标导航
reflective tape
反光带

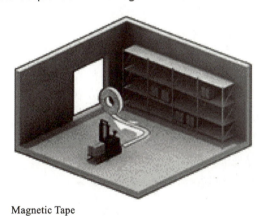

Magnetic Tape

Fig. 14-5 Magnetic Tape Navigation

  Laser target navigation is done by mounting reflective tape on walls, poles, or fixed machines, as shown in Fig. 14-6. The AGV carries a laser transmitter and receiver on a rotating turret. The laser is transmitted and received by the same sensor. The angle and (sometimes) the distance to any reflector that is in the line of sight and in the range are automatically calculated. The information is compared to the map of the reflector layout stored in the AGV's memory and this allows the navigation system to triangulate the current position of the AGV. The current position is compared to the path programmed into the reflector layout map. The steering is adjusted accordingly to keep the AGV on track. The system can then navigate the AGV to a desired target using the constantly updating position.

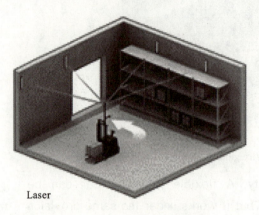

Fig. 14-6 Laser Target Navigation

 New Words and Phrases

portable [ˈpɔːtəbl] *adj*. 便携式的，轻便的
magnet [ˈmæɡnət] *n*. 磁铁，磁石，有吸引力的人（或地方、事物），磁体
laser [ˈleɪzə(r)] *n*. 激光，激光器
laser guided vehicle 激光导引车
navigation [ˌnævɪˈɡeɪʃn] *n*. 导航，领航，航行
laser target navigation 激光目标导航
tow [təʊ] *v*. 牵引，拖，拉，拽
trailer [ˈtreɪlə(r)] *n*. 拖车，挂车
pulp [pʌlp] *n*. 纸浆，浆状物，髓
magnetic [mæɡˈnetɪk] *adj*. 磁的，磁性的
magnetic tape 磁带
destination [ˌdestɪˈneɪʃn] *n*. 目的地，终点
slot [slɒt] *n*. 窄缝，（名单、日程或节目表中的）位置，时间
approximately [əˈprɒksɪmətli] *adv*. 大概，大约，约莫
transmit [trænzˈmɪt] *v*. 输送，发射，传播，传染，使通过
regulate [ˈreɡjuleɪt] *v*. 调节，控制（速度、压力等）
steering [ˈstɪərɪŋ] *n*. （车辆等的）转向装置
circuit [ˈsɜːkɪt] *n*. 线路，电路，环形路线

polarity [pəˈlærəti] n. 极性，截然对立，两极化
dual polarity 双极性
sequence [ˈsiːkwəns] n. 顺序，次序，一系列，一连串
reflective [rɪˈflektɪv] adj. 反射的，反光的
reflective tape 反光带
rotating [rəʊˈteɪtɪŋ] adj. 旋转的，轮值的
turret [ˈtʌrət] n. 塔楼，角楼，炮塔
triangulate [traɪˈæŋɡjuleɪt] v. 对……进行三角测量，由三角形组成的

Notes

1. a variety of 各种各样的

例句：AGCs are available in a variety of models and can be used to move products on an assembly line, transport goods throughout a plant or a warehouse, and deliver loads.

AGC 有多种型号，可用于在流水线上移动产品、在工厂或仓库内运送货物。

例句：He resigned for a variety of reasons.

他由于种种原因辞职了。

2. play an important role in 在……中起重要作用，扮演重要角色

例句：Today, the AGV plays an important role in the design of new factories and warehouses, safely moving goods to the destination.

由于自动导引车可以将货物安全送达目的地，如今它在新型工厂和仓库的设计中发挥着重要作用。

例句：The media play an important role in influencing people's opinions.

媒体在影响舆论方面发挥着重要作用。

3. be fitted with 配有……，装有……

例句：The AGV is fitted with the appropriate guide sensor to follow the path of the tape.

自动导引车配有合适的引导传感器，以跟踪导航条的路径。

例句：Insurance costs will be reduced for houses fitted with window locks.

窗户有锁的房子保险费用会降低。

4. lack sth. 没有，缺乏，不足，短缺

例句：Colored tape is initially less expensive, but it lacks the advantage of being embedded in high-traffic areas where the tape may become damaged or dirty.

彩色导航条购买价格较低，但不能埋入通行量大的区域，因为这些区域的条带可能会破损或污损。

例句：He lacks confidence.

他缺乏信心。

5. be embedded in 被牢牢地埋入(嵌入)，深陷，(态度、感觉等)被深植其中

例句：Colored tape is initially less expensive, but it lacks the advantage of being embedded in high-traffic areas where the tape may become damaged or dirty.

彩色导航条购买价格较低，但不能埋入通行量大的区域，因为这些区域的条带可能会破损或污损。

例句：These attitudes are deeply embedded in our society.

这些看法在我们这个社会中根深蒂固。

 Exercises

Ⅰ. Write True or False beside the following statements about the text.

1. _____ AGVs are most often used in industrial applications to transport heavy materials around a large industrial facility.

2. _____ AGCs cost more than other types of AGVs.

3. _____ LGV is a type of AGV that is equipped with a laser navigation triangulation system.

4. _____ One major advantage of wire guidance over tape guidance is that it can be easily removed and relocated if the course needs to change.

5. _____ Laser target navigation is done by mounting reflective tape on walls, poles, or fixed machines.

Ⅱ. Answer the following questions in English according to the text.

1. What does AGV refer to?
2. What does LGV refer to?
3. How do AGVs move goods to the destination in the wire-guided mode?
4. How do AGVs move goods to the destination in the tape-guided mode?
5. How is laser target navigation implemented?

Ⅲ. Read the text again and fill in the blanks in the following sentences orally.

1. AGVs can _____ objects behind them in _____ to which they can _____ attach. The trailers can be used to move _____ _____ or _____ _____. AGVs are employed in nearly every industry, including pulp, paper, metals, newspaper, and general manufacturing.

2. AGCs are _____ in a _____ of models and can be used to move products on an _____ line, _____ goods throughout a plant or a warehouse, and _____ loads.

3. In an _____ process, LGVs are _____ to communicate with other robots to ensure products are moved _____ through the warehouse, whether they are being stored for future use or sent directly to _____ areas. Today, the AGV plays an _____ role in the design of new factories and warehouses, safely moving goods to the _____.

4. In the wire-guided mode, a sensor is installed on the _____ of the AGV close to the ground. It detects the _____ _____ of the radio signal being _____ from the wire. The information will be used to _____ the _____ circuit, making the AGV follow the wire.

5. Colored tape is initially less _____ , but it _____ the advantage of being _____ in _____ areas where the tape may become damaged or dirty.

Ⅳ. Translation.

Over the years the technology has become more sophisticated and today automated vehicles are mainly laser-navigated, e.g. LGVs (laser guided vehicles). LGV is a type of AGV that is equipped with a laser navigation triangulation system. In an automated process, LGVs are programmed to communicate with other robots to ensure products are moved smoothly through the warehouse, whether they are being stored for future use or sent directly to shipping areas. Today, the AGV plays an important role in the design of new factories and warehouses, safely moving goods to the destination.

Lesson 15   Teach Pendant Manual for Controller-B

The teach pendant (TP) is a portable terminal for operating and controlling the axes connected to controller-B. The TP is equipped with① an EMERGENCY STOP push-button, an AUTO/TEACH selector switch, and a Deadman switch. It can be either hand-held (thus disabling its ability to run programs) or② mounted in a special fixture outside the robot's working envelope (thus enabling access to running programs).

Make sure the teach pendant (or emergency by-pass plug) is properly connected to the controller before you power on the system.

The operation of the teach pendant will vary according to the manner in which it is held.

The Deadman switch and the EMERGENCY STOP push-button ensure the safety of the operator.

1. AUTO/TEACH Selector Switch

When the TP is mounted, the AUTO/TEACH selector switch (as shown in Fig. 15-1) affects control functions and system operation.

2. Deadman Switch

When the hand-held TP is in the TEACH mode, the TP has full

Margin Note

emergency by-pass plug
紧急旁通插头

control of the axes if the Deadman switch (as shown in Fig. 15-2) is depressed. If this switch is released, the system goes into the EMERGENCY state.

3. EMERGENCY STOP Push-button

When the button is pressed, the system goes into the EMERGENCY state (as shown in Fig. 15-3).

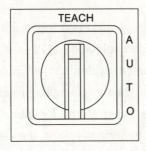
Fig. 15-1 AUTO/TEACH Selector Switch

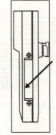
Fig. 15-2 Deadman Switch

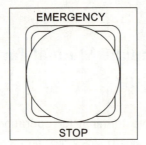
Fig. 15-3 EMERGENCY STOP Push-button

4. Keypad Functions

The teach pendant's keypad (as shown in Fig. 15-4) has 25 coded keys. Most of the keys are multi-functional. Following are descriptions of the teach pendant's keys.

(1) [ENTER/EXECUTE] It accepts and/or executes the command which has been entered and starts the execution of a program.

(2) [JOINTS/XYZ/TOOL] It switches the command mode between JOINTS and Cartesian (XYZ).

Cartesian
笛卡儿

(3) [CLR/GROUP SELECT] It clears a partially entered command and enables TP control of a specific axis group.

(4) [+] In JOINTS mode, it moves the selected axis in a positive direction. In XYZ mode, it moves the tip of the gripper in a positive direction. If group G is selected, it opens the gripper. It confirms the DELETE command.

positive direction
正向

Unit Ⅲ　Manufacturing Technology Informatization | 119

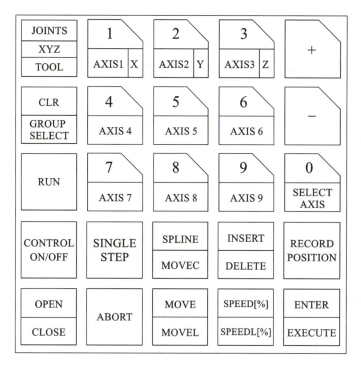

Fig. 15-4　The Keypad of TP

(5) ![-] In JOINTS mode, it moves the selected axis in a negative direction. In XYZ mode, it moves the tip of the gripper in a negative direction. If group G is selected, close the gripper.　　negative direction 负向

(6) ![0/SELECT AXIS] Numerical key 0. It selects axis 1 through 12. (Note: when the number of the axis is over 9, press SELECT AXIS, then press the axis number.)

(7) ![1][2][3] Numerical keys 1, 2, 3, AXIS 1, 2, 3 in JOINTS mode, AXIS X, Y, Z in XYZ mode respectively.

(8) ![4] Numerical key 4, AXIS 4 in JOINTS mode, pitch axis in XYZ mode (MK2 and ER IX only), roll axis in XYZ mode (ER 14 only).　　pitch axis 俯仰轴

(9) ![5] Numerical key 5, AXIS 5 in JOINTS mode, roll axis in XYZ mode (MK2 and ER IX only).　　roll axis 滚动轴

(10) ![6][7][8][9] Numerical keys 6-9, AXIS 6-9 in JOINTS mode.

(11) ![CONTROL ON/OFF] It enables and disables control of the selected group, or all groups.

(12) ![RECORD POSITION] This command both defines and records a position.

(13) ![INSERT/DELETE] This command is used to add and remove positions in a

vector.

（14） ![SPEED] It sets the speed of manual axis movement, as a percentage of maximum speed.

（15） ![OPEN/CLOSE] It opens and closes the gripper. This command functions on both electric and pneumatic grippers.

（16） ![MOVE/MOVEL] It moves the axes to a target position. MOVEL applies only to[3] the robot (group A) axes.

（17） ![SPLINE/MOVEC] It moves the axes to a target position. MOVEC applies only to the robot (group A) axes. SPLINE applies only to multi-axis devices in group A or group B axes.

（18） ![RUN] It executes a program, available only when the TP is mounted.

（19） ![SINGLE STEP] This command is not currently available.

（20） ![ABORT] It aborts the execution of all running programs and stops the robot and all peripheral axes.

5. Display Panel

The teach pendant has a four-line liquid crystal display panel. Each line displays a specific type of message or text.

crystal
display panel
液晶显示面板

6. Messages

When the TP is in either the TEACH mode or AUTO mode, line 1 displays system and error messages.

7. Commands

When the TP is in the TEACH mode, lines 2 and 3 display commands.

8. Status Report

When the TP is in the TEACH mode, line 4 displays the current status of the axes.

New Words and Phrases

terminal ['tɜːmɪnl] n. 终端，终点站，航站楼
portable terminal 便携式终端

push-button 按钮
hand-held 手持式的
fixture [ˈfɪkstʃə(r)] n. 固定装置,夹具
run [rʌn] v. 运行,运转,运作,跑步
depress [dɪˈpres] v. 按,压,推下(尤指机器部件)
keypad [ˈkiːpæd] n. 按键
joint [dʒɔɪnt] n. 关节
confirm [kənˈfɜːm] v. 确定,确认,证实,证明,批准,认可,使坚定,加强
percentage [pəˈsentɪdʒ] n. 百分率,百分比
pneumatic [njuːˈmætɪk] adj. 气动的,压缩空气推动(操作)的,风动的
target [ˈtɑːɡɪt] n. 目标,指标
available [əˈveɪləbl] adj. 可获得的,可找到的,有空的,未婚的
abort [əˈbɔːt] v. 中止,中辍(计划、活动等)
peripheral [pəˈrɪfərəl] adj. 外围的,周边的
erase [ɪˈreɪz] v. 擦掉,抹掉,删除
emergency by-pass plug 紧急旁通插头
Cartesian [kɑːˈtiːziən] 笛卡儿
positive direction 正向
negative direction 负向
pitch axis 俯仰轴
roll axis 滚动轴
crystal display panel 液晶显示面板

 Notes

1. equipped sth. with sth. 配备,装备

例句:The teach pendant is equipped with an EMERGENCY STOP push-button, an AUTO/TEACH selector switch, and a Deadman switch.

该示教器配备紧急停止按钮、自动/示教选择器开关和失能开关。

例句:He equipped himself with a street plan.

他随身带着一张街道平面图。

2. either...or...(对两事物的选择)要么……要么,不是……就是,或者……或者

例句:The teach pendant can be either hand-held (thus disabling its ability to run programs) or mounted in a special fixture outside the robot's working envelope (thus enabling access to running programs).

示教器可以是手持式的(因此无法运行程序),也可以安装在机器人工作包外的专用固定装置中(因此可以访问运行的程序)。

例句：I'm going to buy either a camera or a cellphone with the money.

我打算用这笔钱买一台照相机或者手机。

3. apply to 适用于

例句：MOVEL applies only to the robot（group A）axes.

MOVEL 仅适用于机器人（A 组）轴。

例句：The convention does not apply to us

这条惯例对我们不适用。

 Exercises

Ⅰ. Write True or False beside the following statements about the text.

1. _____ The TP is a terminal that can only be mounted in a fixture.

2. _____ The Deadman switch and the EMERGENCY STOP push-button are for the safety of the operator.

3. _____ If the Deadman switch is depressed, the TP has full control of the axes in TEACH mode.

4. _____ All the keys on the keypad are multi-functional.

5. _____ The teach pendant has a five-line liquid crystal display panel. Each line displays a specific type of message or text.

Ⅱ. Answer the following questions in English according to the text.

1. What is the teach pendant used for?

2. How many keys are there in the TP?

3. Which key can be used to add and remove positions in a vector?

Ⅲ. Read the text again and fill in the blanks in the following sentences orally.

1. The teach pendant can be _____ hand-held _____ mounted in a special fixture.

2. The key "CONTROL ON/OFF" _____ and _____ control of the selected group, or all groups.

3. If you want to move the axes to a target position, you can press the key "_____".

4. _____ applies only to multi-axis devices in group A or group B axes.

5. The key "_____" is used to execute a program.

Ⅳ. Translation.

1. AUTO/TEACH Selector Switch

When the TP is mounted, the AUTO/TEACH selector switch affects control functions and system operation.

2. Deadman Switch

When the hand-held TP is in the TEACH mode, the TP has full control of the axes if the Deadman switch is depressed. If this switch is released, the system goes into the EMERGENCY state.

3. EMERGENCY STOP Push-button

When the button is pressed, the system goes into the EMERGENCY state.

Reading Material 9 The Internet of Things

The internet of things is a technological revolution that represents the future of computing and communications, and its development depends on dynamic technical innovation in a number of important fields, from wireless sensors to nanotechnology.

Firstly, in order to connect everyday objects and devices to large databases and networks — and indeed to the network of networks (the internet) — a simple, unobtrusive, and cost-effective system of item identification is crucial. Only then can data about things be collected and processed. Radio-frequency identification (RFID) offers this functionality. Secondly, data collection will benefit from the ability to detect changes in the physical status of things, using sensor technologies. Embedded intelligence in the things themselves can further enhance the power of the network by devolving information processing capabilities to the edges of the network. Finally, advances in miniaturization and nanotechnology mean that smaller and smaller things will have the ability to interact and connect, as shown in Fig. R9-1.

A combination of all of these developments will create an internet of things that connects the world's objects in both a sensory and an intelligent manner. Indeed, with the benefit of integrated information processing, industrial products, and everyday objects will take on smart characteristics and capabilities. They may also take on electronic identities that can be queried remotely, or be equipped with sensors for detecting physical changes around them. Eventually, even particles as small as dust might be tagged and networked. Such developments will turn the merely static objects of today into newly dynamic things, embedding intelligence in our environment, and stimulating the creation

Margin Note

wireless sensor
无线传感器

nanotechnology
纳米技术

radio-frequency
identification
(RFID)
射频识别

embedded
intelligence
嵌入式智能

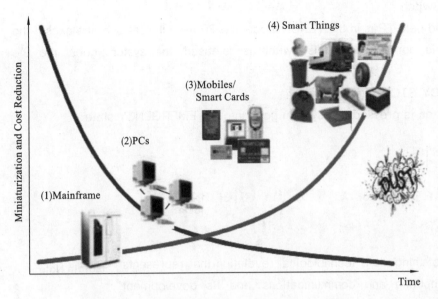

Fig. R9-1 Miniaturization Towards the Internet of Things

of innovative products and entirely new services.

 RFID technology, which uses radio waves to identify items, is seen as one of the pivotal enablers of the internet of things. Although it has sometimes been labeled as the next generation of bar codes, RFID systems offer much more in that they can track items in real-time to yield important information about their location and status. Early applications of RFID include automatic highway toll collection, supply chain management (for large retailers), pharmaceuticals (for the prevention of counterfeiting), and e-health (for patient monitoring). More recent applications range from sports and leisure (ski passes) to personal security (tagging children at schools). RFID tags are even being implanted under human skin for medical purposes, but also for VIP access to bars like the Baja Beach Club in Barcelona. E-government applications such as RFID in driver's licenses, passports, or cash are under consideration. RFID readers are now being embedded in mobile phones.

 In addition to RFID, the ability to detect changes in the physical status of things is also essential for recording changes in the environment. In this regard, sensors play a pivotal role in bridging the gap between the physical and virtual worlds and enabling things to respond to changes in their physical environment. Sensors collect data from their environment, generating information and raising awareness about context. For example, sensors in an electronic jacket can collect

radio wave
无线电波

bar code
条形码

supply-chain
供应链
pharmaceutical
药品
e-health
电子健康

e-government
电子政务
driver's license
驾照

information about changes in external temperature and the parameters of the jacket can be adjusted accordingly.

Embedded intelligence in things themselves will distribute processing power to the edges of the network, offering greater possibilities for data processing and increasing the resilience of the network. This will also empower things and devices at the edges of the network to take independent decisions. "Smart things" are difficult to define, but imply a certain processing power and reaction to external stimuli. Advances in smart homes, smart vehicles, and personal robotics are some of the leading areas. Research on wearable computing (including wearable mobility vehicles) is swiftly progressing.

wearable computing 穿戴式计算

Scientists are using their imagination to develop new devices and appliances, such as intelligent ovens that can be controlled through phones or the internet, online refrigerators, and networked blinds. The internet of things will draw on the functionality offered by all of these technologies to realize the vision of a fully interactive and responsive network environment.

online refrigerator 在线冰箱

New Words and Phrases

dynamic [daɪˈnæmɪk] *adj*. 动态的 *n*. 动力,动力学
nanotechnology [ˌnænəʊtekˈnɒlədʒi] *n*. 纳米技术
database [ˈdeɪtəbeɪs] *n*.(储存在计算机中的)数据库
unobtrusive [ˌʌnəbˈtruːsɪv] *adj*. 不引人注目的,不张扬的,不招摇的
crucial [ˈkruːʃl] *adj*. 关键的,至关重要的,关键性的
devolve [dɪˈvɒlv] *v*. 转移,移交,(使)(权力、职责等)下放
identity [aɪˈdentəti] *n*. 身份,本体,个性,特性
remotely [rɪˈməʊtli] *adv*. 远程地,微弱地
particle [ˈpɑːtɪkl] *n*. 微粒,粒子,颗粒
entirely [ɪnˈtaɪəli] *adv*. 完全地,全部地,完整地
pivotal [ˈpɪvətl] *adj*. 关键性的,核心的
toll [təʊl] *n*. 通行费 *v*. 收费
retailer [ˈriːteɪlə(r)] *n*. 零售商,零售店
counterfeit [ˈkaʊntəfɪt] *v*. 伪造,仿造,制假

security [sɪˈkjʊərəti] n. 安全,保护措施
bridge [brɪdʒ] v. 弥合,桥梁
resilience [rɪˈzɪliəns] n. 恢复力,弹力,适应力
empower [ɪmˈpaʊə(r)] v. 授权,给(某人)……的权力
stimuli [ˈstɪmjʊlaɪ] n. 促进因素,激励因素,刺激物
blind [blaɪnd] n. 窗帘,用以蒙蔽人的言行,借口
wireless sensor 无线传感器
e-health 电子健康
e-government 电子政务
driver's license 驾照
wearable computing 穿戴式计算

Reading Material 10　　Multimedia Technology

Traditionally, computer systems dealt exclusively with numerical calculations. However, text processing soon became an important concern for computer designers. Communications technologies were also developed to support the transmission of textual and numerical data. More recently, there has been a dramatic increase in the range of media types supported by computers and communications technologies. Significant steps have been taken in integrating graphics into computer workstations and communications technology. Researchers are now tackling the harder problems presented by audio and video.

Two themes have emerged from this discussion. Firstly, the variety of media types is an important feature of modern information systems. Secondly, in order to deal with the variety, integration is a critical concern. These observations provide a good working definition of multimedia:

MULTIMEDIA = VARIETY + INTEGRATION

It is necessary for a multimedia system to support a variety of media types. This could be as modest as text and graphics or as rich as animation, audio, and video. However, this alone is not sufficient for a multimedia environment. It is also important that the various sources of media types are integrated into a single system framework, a multimedia system is then one that allows end users to share, communicate and process a variety of forms of information in an

Margin Note

multimedia
多媒体

numerical calculation
数值计算

text processing
文本处理

integrated manner (as shown in Fig. R10-1).

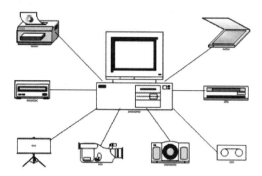

Fig. R10-1 Multimedia Computer System

Today, the various technologies referred to as multimedia define a number of individual niches.

One of the most important of these is animation, the capability to have moving images on your screen. Animation is tightly tied in with another concept called desktop video—actually creating and manipulating video images, to produce in-house presentations, rough drafts of commercial videos, or training products.

desktop video
桌面视频

Sound will also play a key role in a multimedia presentation.

Video images also take up a lot of disk space. To handle this, some groups are looking at optical discs for storage, particularly as erasable optical media became more mainstream.

optical disc
光盘

Desktop video and animation are all well and good, but what many proponents see is a way of combining all these elements into an interactive system—interactive multimedia or hypermedia.

erasable optical media
可擦除光媒体
interactive multimedia
交互式多媒体
hypermedia
超媒体

Attempts to apply computer technology to image, speech, and forecasting share characteristics. All require pattern recognition, or the ability to identify or classify an entity despite noise and distortion. All hold great commercial promises because they reflect the human world. Finally, all are beginning to use neural networks to improve accuracy, reduce cost, or both.

pattern recognition
模式识别
neural network
神经网络

A neural network is an implementation of an algorithm inspired by research into the brain (as shown in Fig. R10-2). In fact, one branch of neuroscience uses computers to model cognitive functions. But the neural networks discussed here have little to do with biology. Rather, they are a technology in which computers learn directly from data, thereby assisting in classification, function estimation, data compression, and similar tasks.

neuroscience
神经科学

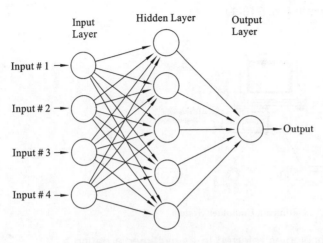

Fig. R10-2 Artificial Neural Networks

Having been used experimentally for decades, neural networks are reputedly a solution in search of a problem. More recently, though, they began moving into practical applications, and this trend can only accelerate now that specialized hardware is available to speed product development. Hundreds of actual applications use the networks, often without public acknowledgment, to preserve a competitive advantage. Products known to be based on them include optical character recognition (OCR) systems, electrocardiographs, process control, and financial systems. Military uses such as target recognition and flight control have also been reported, as well as communications applications like adaptive echo cancellation.

Computer scientists around the world are working to develop neural network software that mimics the human thought process more closely than digital computers do. Full-scale applications of the technology are years away, but they could include enhanced processing of sensory data, and advanced artificial intelligence systems for decision support and robotics, among others.

Because the neural processes are embedded in silicon rather than software, orders-of-magnitude efficiency improvements over existing neural networks are possible.

Neural network programs are generally so large and complex that they require huge memory capacity and are a major processing challenge for even the most powerful computers available today. Here are several ways that researchers are using neural networks:

(1) Pattern recognition   Researchers commonly use filtering operations to solve image recognition and processing problems. These

optical character recognition (OCR) system
光学字符识别系统

electrocardiograph
心电图

adaptive echo cancellation
自适应回声抵消

advanced artificial intelligence system
高级人工智能系统

filtering operation
滤波操作

filtering operations are similar to the processing functions of neurons, so it was natural that the first major application for the neural network was in imaging areas.

(2) Nonlinear processes　Neural networks also are useful for modeling nonlinear processes—those systems where the sum of the inputs is not directly proportional to the output.

nonlinear process
非线性过程

(3) Number crunching　Computation-intensive applications also can be aided by neural network systems.

number crunching
数字运算

New Words and Phrases

exclusively [ɪkˈskluːsɪvli] adv. 唯一地，排他地，独占地
textual [ˈtekstʃuəl] adj. 文本的，篇章的
graphic [ˈɡræfɪk] n. 图表，图形，图画
tackle [ˈtækl] v. 解决，应付，处理
animation [ˌænɪˈmeɪʃn] n. 动画，动画片，活力
audio [ˈɔːdiəʊ] n. 音频，音响
framework [ˈfreɪmwɜːk] n. 框架，（建筑物或物体的）构架
niche [niːʃ] n. 市场定位，生态位
draft [drɑːft] n. 草稿，草案，草图
mainstream [ˈmeɪnstriːm] adj. 主流的
distortion [dɪˈstɔːʃn] n. 失真，扭曲，歪曲
branch [brɑːntʃ] n. 分支，树枝，分部
estimation [ˌestɪˈmeɪʃn] n. 估计，判断，评价
experimentally [ɪkˌsperəˈmentli] adv. 用实验方法地，实验式地
acknowledgment [əkˈnɒlɪdʒmənt] n. 承认，致谢
mimic [ˈmɪmɪk] v. 模仿（人的言行举止），模拟
magnitude [ˈmæɡnɪtjuːd] n. 巨大，重大
neuron [ˈnjʊərɒn] n. 神经元
proportional [prəˈpɔːʃənl] adj. 相称的，成比例的
optical disc 光盘
erasable optical media 可擦除光媒体
interactive multimedia 交互式多媒体
hypermedia [ˌhaɪpəˈmiːdiə] n. 超媒体
pattern recognition 模式识别

optical character recognition (OCR) system 光学字符识别系统
electrocardiograph [ɪˌlektrəʊˈkɑːdɪəɡrɑːf] n. 心电图
adaptive echo cancellation 自适应回声抵消
advanced artificial intelligence system 高级人工智能系统
filtering operation 滤波操作
nonlinear process 非线性过程
number crunching 数字运算

Reading Material 11   Global Positioning System (GPS)

GPS (as shown in Fig. R11-1) is the global positioning system. GPS uses satellite technology to enable a terrestrial terminal to determine its position on the earth in latitude and longitude.

Margin Note

global positioning system(GPS)
全球定位系统
terrestrial terminal
地面终端
latitude
纬度
longitude
经度

Fig. R11-1   GPS

GPS receivers do this by measuring the signals from three or more satellites simultaneously and determining their position using the timing of these signals.

GPS operates using trilateration. Trilateration is the process of determining the position of an unknown point by measuring the lengths of the sides of an imaginary triangle between the unknown point and two or more known points.

trilateration
三边测量

In the GPS, the two known points are provided by two GPS satellites. These satellites constantly transmit an identifying signal. The GPS receiver measures the distance to each GPS satellite by measuring the time each signal took to travel between the GPS satellite and the GPS receiver.

The GPS is divided into three segments:

(1) The space segment.
(2) The control segment.
(3) The user segment.

GPS uses twenty-one operational satellites, with an additional three satellites in orbit as a redundant backup. GPS uses NAVSTAR satellites manufactured by Rockwell International. Each NAVSTAR satellite is approximately 5 meters wide (with solar panels extended) and weighs approximately 900 kg.

GPS satellites orbit the earth at an altitude of approximately 20 200 km. Each GPS satellite has an orbital period of 11 hours and 58 minutes. This means that each GPS satellite orbits the earth twice each day.

These twenty-four satellites orbit in six orbital planes, or paths. This means that four GPS satellites operate in each orbital plane. Each of these six orbital planes is spaced sixty degrees apart. All of these orbital planes are inclined fifty-five degrees from the equator.

For GPS tracking to work, it is necessary to have both access to the global positioning system and a GPS receiver (as shown in Fig. R11-2). The GPS receiver can receive signals that are transmitted by GPS satellites orbiting overhead. Once these satellite transmissions are received by the GPS receiver, location and other information such as speed and direction can be calculated.

Fig. R11-2 A GPS Receiver

The receiver contains a mathematical model to account for these influences, and the satellites also broadcast some related information which helps the receiver in estimating the correct speed of propagation. Certain delay sources, such as the ionosphere, affect the speed of radio waves based on their frequencies, dual-frequency receivers can actually measure the effects on the signals.

In order to measure the time delay between the satellite and receiver, the satellite sends a repeating 1 023-bit long pseudo-random sequence; the receiver constructs an identical sequence and shifts it until the two sequences match.

Different satellites use different sequences, which lets them all broadcast on the same frequencies while still allowing receivers to distinguish between satellites. This is an application of code division multiple access (CDMA).

code division multiple access (CDMA)
码分多址

There are two frequencies in use: 1 575.42 MHz (referred to as L1), and 1 227.60 MHz (L2). The L1 signal carries a publicly usable coarse-acquisition (C/A) code as well as an encrypted P(Y) code. The L2 signal usually carries only the P(Y) code.

coarse-acquisition (C/A) code
粗捕获码
encrypted P(Y) code
加密码

New Words and Phrases

satellite [ˈsætəlaɪt] n. 人造卫星,卫星
latitude [ˈlætɪtjuːd] n. 纬度,纬度地区
longitude [ˈlɒŋgɪtjuːd] n. 经度
trilateration [ˌtraɪlætəˈreɪʃən] n. 三边测量
segment [ˈsegmənt, segˈment] n. 部分,份,片,段
orbit [ˈɔːbɪt] n. (天体等运行的)轨道,影响范围,势力范围
redundant [rɪˈdʌndənt] adj. 冗余的,被裁减的,不需要的
weigh [weɪ] v. 重,称重
propagation [ˌprɒpəˈgeɪʃ(ə)n] n. 传播,扩展,宣传
pseudo [ˈsjuːdəʊ] adj. 假的,冒充的
encrypted [ɪnˈkrɪptɪd] adj. 加密的
global positioning system (GPS) 全球定位系统
terrestrial terminal 地面终端
operational satellite 运行卫星
redundant backup 冗余备份
solar panels extended 太阳能电池板延伸
orbital plane 轨道平面
equator [ɪˈkweɪtə(r)] n. 赤道
ionosphere [aɪˈɒnəsfɪə(r)] n. 电离层
dual-frequency receiver 双频接收器

code division multiple access（CDMA）码分多址
coarse-acquisition（C/A）code 粗捕获码
encrypted P(Y) code 加密码

Reading Material 12   Virtual Instruments

Virtual instruments (as shown in Fig. R12-1) are computer programs that interact with real-world objects by means of sensors and implement functions of real or imaginary instruments. The sensor is usually a simple hardware that acquires data from the object, transforms it into electric signals, and transmits it to the computer for further processing. Simple virtual measuring instruments just acquire and analyze data, but more complex virtual instruments communicate with objects in both directions.

Margin Note

virtual instrument
虚拟仪器

Fig. R12-1   Virtual Instrument

Real-world signals are of analog nature, while a computer is a digital instrument; therefore the computer needs also interpreters, analog-to-digital and digital-to-analog converters for communication with the object. ADC and DAC boards that implement this function in inexpensive systems are usually placed inside the computer. Compact external ADC/DAC converters with USB interfaces are also becoming popular.

LabVIEW is a graphical programming language from National Instruments. LabVIEW programs are called virtual instruments often abbreviated to VIs. Each virtual instrument has two components: a block diagram, and a front panel. Controls and indicators on the front panel allow an operator to input data into or extract data from an already-running virtual instrument.

analog-to-digital
converter(ADC)
模数转换

digital-to-analog
converter(DAC)
数模转换

block diagram
框图

front panel
前面板

In terms of performance, LabVIEW includes an actual compiler that produces native code for the CPU platform, so the graphical code is compiled, rather than interpreted.

One of the main benefits of LabVIEW is that people with little or no previous programming experience are able to access hardware input/output more rapidly and through a hardware abstraction system. This abstraction allows isolation between hardware implementation and software solution. A technique without National Instruments software driver interface would be extremely time-consuming.

Another virtual instrumentation component is modular I/O, designed to be rapidly combined in any order or quantity to ensure that virtual instrumentation can both monitor and control any development aspect. Using well-designed software drivers for modular I/O, engineers and scientists quickly can access functions during concurrent operation.

The third virtual instrumentation element — using commercial platforms, often enhanced with accurate synchronization — ensures that virtual instrumentation takes advantage of the very latest computer capabilities and data transfer technologies. This element delivers virtual instrumentation on a long-term technology base that scales with the high investments made in processors, buses, and more.

Virtual instrumentation systems frequently use ethernet for remote test system control, distributed I/O, and enterprise data sharing. Ethernet provides a low-cost, moderate-throughput method for exchanging data and controlling commands over distances. However, due to its packet-based architecture, ethernet is not deterministic and has relatively high latency. For some applications, such as instrumentation systems, the lack of determinism and high latency make ethernet a poor choice for integrating adjacent I/O modules. These situations are better served with a dedicated bus such as PXI, VXI, or GPIB(as shown in Fig. R12-2).

native code
本地代码
CPU
中央处理器

hardware abstraction system
硬件抽象系统

concurrent operation
并行操作系统
synchronization
同步性

ethernet
以太网

moderate-throughput
中等吞吐量

dedicated bus
专用总线

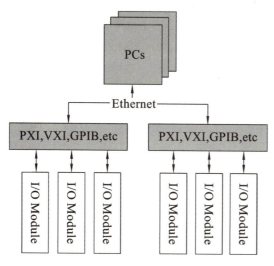

Fig. R12-2　Virtual Instrumentation Systems are Served with a Dedicated Bus

 New Words and Phrases

imaginary [ɪˈmædʒɪnəri] *adj*. 想象中的,幻想的,虚构的
analog [ˈænəlɒɡ] *adj*. 模拟的,指针式的
instrument [ˈɪnstrəmənt] *n*. 器械,仪器,器具
interpreter [ɪnˈtɜːprɪtə] *n*. 读卡机,解释程序,口译译员,演绎者,表演者
converter [kənˈvɜːtə] *n*. 转换器,使发生转化的人(或物)
graphical [ˈɡræfɪkl] *adj*. 图形的,用图(或图表等)表示的
programming [ˈprəʊɡræmɪŋ] *n*. (计算机)程序设计,程序编制,编程
enhance [ɪnˈhɑːns] *v*. 增强,提高,改善
abbreviate [əˈbriːvieɪt] *v*. 缩略,缩写
compile [kəmˈpaɪl] *v*. 编写,编纂
compiler [kəmˈpaɪlə(r)] *n*. 编纂者,汇编者
previous [ˈpriːviəs] *adj*. 以前的,先前的,以往的
modular [ˈmɒdjələ(r)] *adj*. 模块化的,组合式的
quantity [ˈkwɒntəti] *n*. 量,数量,大量
concurrent [kənˈkʌrənt] *adj*. 同时发生的,并存的
platform [ˈplætfɔːm] *n*. 平台,月台,讲台
moderate [ˈmɒdəreɪt] *adj*. 适度的,中等的,温和的
packet [ˈpækɪt] *n*. 小包裹,(商品的)小包装纸袋,小硬纸板盒
deterministic [dɪˌtɜːmɪˈnɪstɪk] *adj*. 基于决定论的,不可抗拒的,不可逆转的

adjacent [əˈdʒeɪsnt] *adj.* 相邻的，邻近的，与……毗连的
virtual instrument 虚拟仪器
analog-to-digital converter（ADC）模数转换
digital-to-analog converter（DAC）数模转换
block diagram 框图
native code 本地代码
CPU 中央处理器
hardware abstraction system 硬件抽象系统
concurrent operation 并行操作系统
moderate-throughput 中等吞吐量
ethernet [ˈiːθənet] *n.* 以太网
dedicated bus 专用总线

Unit Ⅳ Modern Production Management Technology

Introduction to the Unit 单元导言

> 智能制造贯穿于设计、生产、管理、服务等制造活动的各个环节。本单元将重点讨论有关现代制造管理技术方面的内容，主要是资产状态监控、制造执行系统、产品数据管理、制造资源计划、质量控制、企业资源计划等内容以及与生产一线相关的计算机辅助工艺设计、智能制造控制制造执行系统软件、焊接缺陷自动检测等内容。本单元的目的是让学生了解现代生产控制目标，理解最大的客户服务、最小的库存投资、高效率生产在智能制造中的重要性，初步掌握现代生产管理中的英文表达。

Lesson 16 Asset Status Monitoring

Asset monitor (AM) gives you detailed real-time and historical information about what is happening on the shop floor. Graphs, charts, and summary reports are an important part of asset monitor, but if the data behind these presentations is inaccurate or unreliable, then even the most sophisticated presentations are of little value. The accuracy and reliability of data gathered from machine tools vary widely, depending on the equipment and methods for data collection. Only with advanced CNC communication technology and integration of physical and logical data access capabilities can we ensure that the data obtained from the machine tool is accurate and meaningful.

The system allows the flexibility to monitor feed hold, feed speed coverage, spindle, shaft motion, coolant on/off, and much more, accurate measurement of production (as shown in Fig. 16-1). Combine each machine signal to get a true indication of when the machine "makes the chip". People are tired of scrolling through① a lengthy list of the report to get a piece of information. The intuitive user interface of the asset monitor provides a detailed view of shop production quickly and easily. Just one or two clicks of the mouse and you have the information you

Margin Note

shop floor
车间

machine tools
机床

feed hold
持续给进
shaft motion
传动轴

need.

Fig. 16-1　Asset Monitor System

1. Features and Functions of Asset Monitor System

It offers real-time monitoring and provides production statistics, overall equipment effectiveness (OEE), mobile phone texting, and email notifications on key machine signals and events. The system archives your production data so that you can historically view shop floor performance over a year, a month, a week, a day, or a shift period.

2. Derived Signals

Asset monitor allows users to combine machines' signals to create a higher-level derived signal. For example, the derived signal, "Cutting Part", can be created to represent the following: Power On + Cycle On + Spindle On + Feed Hold Off + Axis Motion + Feed Override ≥ 80%. When the above presentation is true, the derived signal "Cutting" is considered to be on.

derived signal
衍生信号

feed override
速度倍率设定

3. Charts and Reports to Fit Your Needs

Once you have configured the system to monitor the machine signals that are important to the operations, asset monitor provides an easy-to-use, intuitive web-based utilization and performance viewer, as shown in Fig. 16-2.

4. Supported CNC Controls and Signals

Asset monitor is a flexible system providing ethernet connectivity to all equipment. A direct ethernet connection can be made to CNC controls supporting the following interfaces.

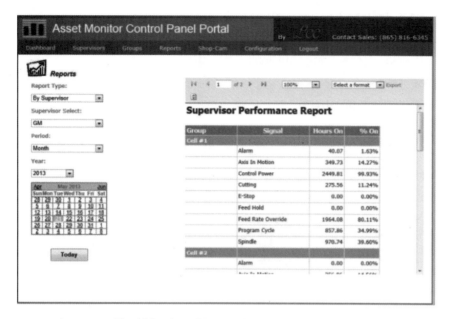

Fig. 16-2　Asset Monitor Control Panel Portal

(1) Control power on/off.

(2) Spindle on/off, actual spindle speed.　　　spindle speed 主轴转速

(3) Program cycle on/off, running program.

(4) Feed hold on/off.

(5) Feed override.

(6) Spindle override.

(7) Axis in motion, actual feed rate.

(8) E-stop.　　　e-stop 紧急停工

(9) CNC alarm with message.

(10) Coolant on/off, etc.

5. Part Count Supported

Get an accurate count on the number of good and bad parts produced in your facility. Asset monitor keeps track of[②] the number of parts machined if they are good or bad, and the CNC program used to machine them. Based on this data, asset monitor calculates the OEE quality metric over a year, month, week, day, or shift. Asset monitor also breaks down[③] the number of parts machined per hour on a given date.

CNC program 计算机数控程序

6. Text Message and E-mail Notifications

Configure your system to send text messages and/or e-mail notifications when a specific machine event occurs (i.e. alarmed or out of cycle for some time).

add-on utility 附加实用程序

This application is designed as an add-on utility for the asset monitor system. It can be deployed on large flat panel monitors out on the shop floor or on an office PC to help foremen and supervisors quickly identify maintenance issues, machine faults, downtime, out of stock, and performance issues.

downtime
停工时间

7. Cost Effective and Scalable

Asset monitor is designed to fit a wide variety of④ implementations. From small shop environments supporting several machines to the wide multi-facility solutions of enterprises, asset monitor can easily be scaled to⑤ fit your needs⑥.

Tips

> **1. OEE: Overall Equipment Effectiveness 设备综合效率**
>
> 一般地，每一个生产设备都有自己的理论产能，要实现这一理论产能就必须保证没有任何干扰和质量损耗。OEE 表示实际的生产能力相对于理论产能的比率。它用来评价停机所带来的损失，包括引起计划生产停工的任何事件，例如设备故障、原材料短缺以及生产方法的改变等。

New Words and Phrases

advanced [əd'vɑːnst] *adj.* 先进的，高级的，高等的
flexibility [ˌfleksə'bɪləti] *n.* 灵活性，弹性，柔性
spindle ['spɪnd(ə)l] *n.* 轴，细长的人或物
coolant ['kuːlənt] *n.* 冷却剂
indication [ˌɪndɪ'keɪʃn] *n.* 指示，象征，迹象
scroll [skrəʊl] *v.* 滚屏，滚动，使相纸卷合（或打开）那样移动
notification [ˌnəʊtɪfɪ'keɪʃn] *n.* 通知，通告，布告
archives ['ɑːkaɪvz] *n.* 档案，案卷（archive 的复数）
shift [ʃɪft] *n.* 轮班，工作时间改变，转变 *v.* （使）移动，（使）转移
derived [dɪ'raɪvd] *adj.* 导出的，衍生的，派生的
derived signal 衍生信号
override [ˌəʊvə'raɪd] *n.* 超控装置，预算超量，增加
feed override 速度倍率设定

configure [kənˈfɪɡə(r)] v. (计算机)配置,安装,设定
calculate [ˈkælkjuleɪt] v. 计算,核算,预测,推测
metric [ˈmetrɪk] n. 指标,衡量标准,度规
foreman [ˈfɔːmən] n. 领班,陪审团主席
scalable [ˈskeɪləbl] adj. 可扩展的,可去鳞的,可称量的
shop floor 车间
feed hold 持续给进
shaft [ʃɑːft] n. 轴,杆,柄,箭
shaft motion 传动轴
spindle speed 主轴转速
e-stop 紧急停工
CNC program 计算机数控程序
utility [juːˈtɪləti] n. 应用程序,实用程序
add-on utility 附加实用程序

 Notes

1. scroll through 浏览,滚动,翻阅

例句:People are tired of scrolling through a lengthy list of the report to get a piece of information.

人们厌倦了为了一条信息去翻阅冗长的一系列报告。

例句:Now, we can scroll through our data offline.

现在,我们可以离线浏览数据。

2. keep track of 记录,与……保持联系

例句:Asset monitor keeps track of the number of parts machined if they are good or bad, and the CNC program used to machine them.

资产监控可跟踪加工的零件数量(无论零件是合格零件还是次品)以及用于加工它们的计算机数控程序。

例句:It was a very effective way for them to keep track of their spending.

对他们来说,这是一种非常有效的记录支出的方法。

3. break down 分解,破

例句:Asset monitor also breaks down the number of parts machined per hour on a given date.

资产监控还细分了给定日期内每小时加工的零件数量。

例句:She had been waiting for Simon to break down the barrier between them.

她一直在等着西蒙来破除他们之间的隔阂。

4. a wide variety of 多种的，各种各样的

例句：Asset monitor is designed to fit a wide variety of implementations.
资产监控被设计适用于各种场合。

例句：I'm sure we will find a wide variety of choices available in school cafeterias.
我相信我们会在学校的自助餐厅里找到各种各样的选择。

5. be scaled to 被调整至……，被缩放至……

例句：Asset monitor can easily be scaled to fit your needs.
资产监控可很容易地被扩展以满足各种需求。

例句：Most importantly, the flexible design can be scaled to fit almost any location.
最重要的是，这次改造设计灵活，适用于任何地点。

6. fit one's needs 满足某人的需求

例句：Asset monitor can easily be scaled to fit your needs.
资产监控可很容易地被扩展以满足各种需求。

例句：We can arrange the schedule to fit your needs.
我们可根据您的需求来安排日程。

 Exercises

Ⅰ. Write True or False beside the following statements about the text.

1. _____ Asset monitor (AM) gives you detailed real-time and historical information about what is happening on the shop floor.

2. _____ The accuracy and reliability of data gathered from machine tools vary little, depending on the equipment and methods for data collection.

3. _____ Asset monitor allows users to combine machines, signals to create a higher-level derived signal.

4. _____ Once you have configured the system to monitor the machine signals that are important to the operations, asset monitor provides an easy-to-use, intuitive web-based utilization, and performance viewer.

5. _____ Asset monitor is designed to fit a wide variety of implementations. From small shop environments supporting several machines to the wide multi-facility solutions of enterprises, asset monitor can hardly be scaled to fit your needs.

Ⅱ. Answer the following questions in English according to the text.

1. What does asset monitor refer to? What are the features and functions of the asset monitor system?

2. What are the interfaces of the CNC controls that can be connected to ethernet?

3. What purposes are text messages and e-mail notifications designed as an add-on utility for?

III. Read the text again and fill in the blanks in the following sentences orally.

1. Asset monitor (AM) gives you detailed _____ and _____ information about what is happening on the shop floor.

2. Only with advanced _____ _____ technology and integration of _____ and _____ data access capabilities can we ensure that the data obtained from the machine tool is accurate and meaningful.

3. Features and functions of asset monitor system are offering real-time monitoring and providing _____ _____, overall equipment effectiveness (OEE), mobile phone texting, and _____ _____ on key machine signals and events.

4. Once you have configured the system to monitor the machine signals that are important to the operations, asset monitor provides an easy-to-use, intuitive web-based _____ and _____ _____.

5. Get an accurate count on the number of good and bad parts produced in your facility. Asset monitor keeps track of the number of _____ _____ if they are good or bad, and the CNC program used to machine them.

IV. Translation.

2. Derived Signals

Asset monitor allows users to combine machines' signals to create a higher-level derived signal. For example, the derived signal, "Cutting Part", can be created to represent the following: Power On + Cycle On + Spindle On + Feed Hold Off + Axis Motion + Feed Override ≥ 80%. When the above presentation is true, the derived signal "Cutting" is considered to be on.

Lesson 17 Manufacturing Execution System

Manufacturing execution systems (MESs) are computerized systems used in manufacturing to track and document the transformation of raw materials to finished goods, as shown in Fig. 17-1. MES provides information that helps manufacturing decision-makers understand how current conditions on the plant floor can be optimized to improve production output. MES works as a real-time monitoring system to enable the control of multiple elements of the production process (e.g. inputs, personnel, machines, and support services).

MES can operate across multiple function areas, for example, management of product definitions across the product life cycle, resource scheduling, order execution and dispatch, production analysis

Margin Note

raw material
原材料

monitoring system
监控系统

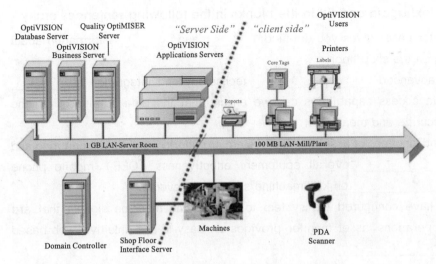

Fig. 17-1　Manufacturing Execution Systems (MES)

and downtime management for overall equipment effectiveness (OEE), product quality, or materials track. MES creates the "as-built" record, capturing the data, processes, and outcomes of the manufacturing process. This can be especially important in regulated industries, such as food and beverage or pharmaceutical, where documentation and proof of processes, events, and actions may be required.

downtime management
停机管理
as-build record
竣工记录

As shown in Fig. 17-2, the idea of MES might be seen as an intermediate step between, on the one hand, an enterprise resource planning (ERP) system, and a supervisory control and data acquisition (SCADA) or process control system on the other; although historically, exact boundaries have fluctuated.

"Manufacturing execution systems help to create flawless manufacturing processes and provide real-time feedback of requirement changes" and provide information from a single source. Other benefits from a successful MES implementation include:

real-time feedback
实时反馈

(1) Reduced waste, re-work, and scrap, including shortened installation time.

(2) More accurate capture of cost information (e. g. labor, scrap, downtime, and equipment).

(3) Increased uptime.

uptime
运行时间

(4) Incorporated paperless workflow activities.

(5) Manufacturing operations traceability.

(6) Decreased downtime and easy-to-detect faults.

(7) Reduced inventory through the eradication of just-in-case inventory.

just-in-case inventory
即时库存

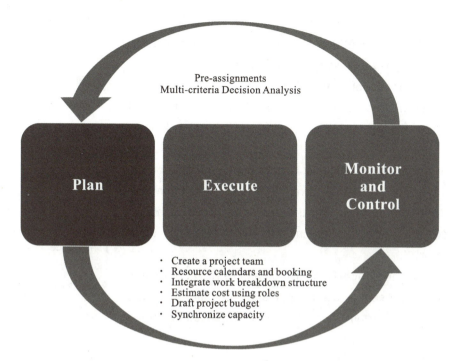

Fig. 17-2　MES May Operate Across Multiple Function Areas

Over the years, international standards and models have refined the scope of such systems in terms of① activities. These typically include:

(1) Management of product definitions. This may include storage, version control, and exchange with other systems of master data like product production rules, bill of material, bill of resources, the process set points, and recipe data, all focusing on② defining how to make a product. Management of product definitions can be part of③ product life cycle management.

bill of material
物料清单

(2) Management of resources. This may include registration, exchange, and analysis of resource information, aiming to④ prepare and execute production orders with resources of the right capabilities and availability.

(3) Scheduling (production processes). These activities determine the production schedule as a collection of work orders to meet the production requirements, typically received from enterprise resource planning (ERP) or specialized advanced planning and scheduling systems, making optimal use of local resources.

(4) Dispatching production orders. Depending on⑤ the types of production processes, it includes further distribution of batches, runs, and work orders, issuing them to work centers, and making adjustments to unanticipated conditions.

(5) Execution of production orders. Although actual execution is done by process control systems, MES may perform checks on resources and inform other systems about the progress of production processes.

(6) Collection of production data. This includes the collection, storage, and exchange of process data, equipment status, material lot information, and production logs in either a data historian or⑥ relational database.

production log
生产日志

(7) Production performance analysis. Create useful information out of⑦ the collected raw data about the current status of production, like work in progress (WIP) overviews, and the production performance of the past period like the overall equipment effectiveness or any other performance indicator.

performance indicator
绩效指标

(8) Production track and trace. Registration and retrieval of related information in order to present a complete history of lots, orders, or equipment (particularly important in health-related productions, e.g. pharmaceuticals).

 Tips

1. ERP: Enterprise Resource Planning 企业资源计划

企业资源计划(ERP)主要用于改善企业业务流程以提高企业核心竞争力。ERP系统是指建立在信息技术基础上,以系统化的管理思想为企业决策层及员工提供决策运行手段的管理平台。

ERP系统包括以下主要功能:供应链管理(SCM)、销售与市场、分销、客户服务、财务管理、制造管理、库存管理、工厂与设备维护、人力资源、报表、制造执行系统(manufacturing execution system, MES)、工作流服务和企业信息系统等。其还包括金融投资管理、质量管理、运输管理、项目管理、法规与标准和过程控制等补充功能。

2. SCADA: Supervisory Control and Data Acquisition 监控与数据采集

监控与数据采集(SCADA)系统是以计算机为基础的分散控制系统(distributed control system, DCS)与电力自动化监控系统。它应用领域很广,可以用于电力、冶金、石油、化工、燃气、铁路等领域的数据采集、监视控制以及过程控制等。SCADA系统主要由以下部分组成:监控计算机、远程终端单元(RTU)、可编程逻辑控制器(PLC)、通信基础设施、人机界面(HMI)。

New Words and Phrases

dispatch [dɪˈspætʃ] v. 调度,发送,迅速处理
intermediate [ˌɪntəˈmiːdiət] adj. 中间的,居中的,中级的
uptime [ˈʌptaɪm] n. 运行时间,上线时间
fluctuate [ˈflʌktʃueɪt] v. 波动,起伏不定,使波动
supervision [ˌsuːpəˈvɪʒ(ə)n] n. 监督,管理
eradication [ɪˌrædɪˈkeɪʃn] n. 消除,根除,消灭
registration [ˌredʒɪˈstreɪʃn] n. 登记,注册,挂号
specialized [ˈspeʃəlaɪzd] adj. 专门的,专业的
lot [lɒt] n. 批次,小块土地,电影摄制场
retrieval [rɪˈtriːv(ə)l] n. 检索,恢复,挽回,找回,取回
monitoring system 监控系统
downtime management 停机管理
as-build record 竣工记录
real-time feedback 实时反馈
just-in-case inventory 即时库存
bill of material 物料清单
production log 生产日志
indicator [ˈɪndɪkeɪtə(r)] n. 标志,迹象,指标
performance indicator 绩效指标

 Notes

1. in terms of 在……方面,依据

例句:Over the years, international standards and models have refined the scope of such systems in terms of activities.
多年来,国际标准和模式在活动方面完善了下列系统的范围。
例句:The data is limited in terms of both quality and quantity.
这份资料在质量和数量上都很有限。

2. focus on 集中(注意力,精神)于,专注于

例句:This all focus on defining how to make a product.
这些都集中于定义如何制造产品。
例句:The talks will focus on the economic development of the region.
会谈将着重讨论该地区的经济发展。

3. be part of 成为……的一部分

例句:Management of product definitions can be part of product life cycle management.

产品定义管理可以是产品生命周期管理的一部分。

例句：Physical exercise should be part of ordinary life.

体育锻炼理应是日常生活的一部分。

4. aim to 计划，以……为目标

例句：This may include registration, exchange, and analysis of resource information, aiming to prepare and execute production orders with resources of the right capabilities and availability.

这包括资源信息的登记、交换和分析，旨在以可利用及能够使用的资源来准备和执行生产订单。

例句：We aim to be there around six.

我们力争六点钟左右到那里。

5. depend on 根据，依赖

例句：Depending on the types of production processes, it includes further distribution of batches, runs, and work orders, issuing these to work centers, and making adjustments to unanticipated conditions.

根据生产过程的类型，这包括进一步分配批次、运行和工作通知单，将其发送给工作中心，并根据意外情况进行调整。

例句：Can you depend on her version of what happened?

你能相信她对所发生事情的描述吗？

6. either...or.... 二者择一的，要么……要么……

例句：This includes the collection, storage, and exchange of process data, equipment status, material lot information, and production logs in either a data historian or relational database.

这包括在历史数据库或关系数据库中收集、存储和交换工艺数据、设备状态、物料批次信息和生产日志。

例句：He didn't want to go on the record as either praising or criticizing the proposal.

他不想公开赞扬或批评这项提议。

7. out of 用……（材料），在……范围（或限制）外

例句：Create useful information out of the collected raw data about the current status of production.

从收集的原始数据中创建有关当前生产状态的有用信息。

例句：I asked out of curiosity.

我因为好奇问了问。

 *Exercises*

Ⅰ. Write True or False beside the following statements about the text.

1. _____ MES provides information that helps manufacturing decision-makers understand how current conditions on the plant floor can be optimized to improve production output.

2. _____ MES creates the "as-built" record, capturing the data, processes, and outcomes of the manufacturing process.

3. _____ MES can't provide real-time feedback on requirement changes.

4. _____ Other benefits from a successful MES implementation might include manufacturing operations traceability.

5. _____ Management of product definitions may include storage, version control, and exchange with other systems of master data.

Ⅱ. Answer the following questions in English according to the text.

1. What does manufacturing execution systems (MES) refer to? What are the main benefits for manufacturing decision-makers to use MES?

2. Which areas does MES may operate in?

3. How does MES do the production performance analysis?

Ⅲ. Read the text again and fill in the blanks in the following sentences orally.

1. Manufacturing execution systems (MESs) are _____ _____ used in manufacturing to track and document the transformation of raw materials to finished goods.

2. "Manufacturing execution systems help to create _____ _____ processes and provide _____ _____ of requirement changes" and provide information from a single source.

3. This may include registration, exchange, and analysis of resource information, aiming to prepare and execute production orders with resources of the right _____ and _____.

4. MES creates the "as-built" record, capturing the data, _____, and _____ of the manufacturing process.

5. Create useful information out of the collected raw data about the current status of production, like work in progress (WIP) overviews, and the production performance of the past period like the _____ _____ effectiveness or any other performance _____.

Ⅳ. Translation.

Management of product definitions. This may include storage, version control, and exchange with other systems of master data like product production rules, bill of material, bill of resources, the process set points, and recipe data, all focusing on defining how to make a product. Management of product definitions can be part of product life cycle management.

Lesson 18　Computer-aided Process Planning

Computer-aided process planning (CAPP) is the use of computer technology to aid in the process planning of a part or product in manufacturing. CAPP is the link between CAD and CAM in that it provides for the planning of the process to be used in producing a designed part.

Process planning is concerned with determining the sequence of individual manufacturing operations needed to produce a given part or product. The resulting operation sequence is documented on a form typically referred to as a "route sheet" (also known as process sheet as shown in Fig. 18-1) containing a list of the production operations and associated machine tools for a work part or assembly. Process planning in manufacturing also refers to the planning of the use of blanks, spare parts, packaging material, user instructions (manuals), etc.

Margin Note

computer-aided
process planning
计算机辅助工艺
设计

route sheet
工艺图表

spare part
备件

Fig. 18-1　A Route Sheet

As the design process is supported by many computer-aided tools, computer-aided process planning (CAPP) has evolved to[①] simplify and improve process planning to achieve more effective use of manufacturing resources.

Computer-aided process planning (CAPP) also includes specifying process routes such as operation methods, operation sequences, workpiece centers, process standards, tools, and fixtures.

Computer-aided process planning evolved as a means to

manufacturing
resources
制造资源

workpiece center
工件中心

electronically store a process plan once it was created, retrieve it, modify it for a new part, and print the plan.

As shown in Fig. 18-2, the process routes used to produce gear shapes generally include milling, pulling, sawing, and filing.

milling
铣削
pulling
拉销
sawing
锯切
filing
锉削

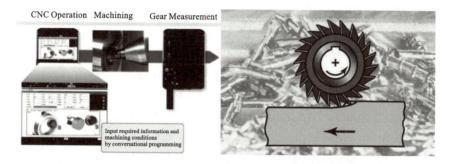

Fig. 18-2 The Process Routes of a Gear Machining

Generative (or dynamic) CAPP is the main focus of development, the ability to automatically generate production plans for new products, or dynamically update production plans on the basis of resource availability. Generative CAPP uses iterative methods, where simple production plans are applied to② automatic CAD/CAM development to refine the initial production plan. Generative CAPP systems are built on③ a factory's production capabilities and capacities. Because generative CAPP systems are based on the unique capabilities and capacity required to produce a given product, each process plan can be defined automatically, independent of④ past process routes. As improvements are made to production efficiencies, the improvements are automatically incorporated into the current product mix. This generative system is a key component of the CAPP system for the agile manufacturing environment.

product mix
产品结构
agile manufacturing environment
敏捷制造环境

In order to achieve the generative CAPP system, components were built to meet the needed capabilities:

(1) The combination of the feature extensions of the process plan and the parametric data associated with⑤ them became part of the data that is passed from the CAD system to the modified product data management (PDM) system as the data set content for the specific product, assembly, or part.

(2) Create a manufacturing execution system (MES). The MES's major component is an expert/artificial intelligent system that matches the engineering feature objects from the product data management (PDM) system with the tooling, personnel, material, transportation

needs, etc. needed to manufacture them in the enterprise resource planning (ERP) system. Once the above physical components are identified, the process items can be scheduled.

(3) The scheduling is continuously updated based on the real-time conditions of the enterprise. Ultimately, the parameters for this system were based on⑥ expenditures, time, physical dimensions, and availability. The parameters are used to produce multidimensional differential equations. Solving the partial differential equations will produce the optimum process and production planning at the time when the solution was generated. Solutions had the flexibility to change over time based on the ability to satisfy agile manufacturing criteria. Execution planning can be dynamic and accommodate changing conditions.

physical dimension
物理尺寸
differential equation
微分方程

New Words and Phrases

blank ['blæŋk] n. 坯料，空白处，空格 adj. 单调的，彻底的
specify ['spesɪfaɪ] v. 明确指出，把……列入说明书
retrieve [rɪ'triːv] v. 找回，收回，检索
modify ['mɒdɪfaɪ] v. 修改，修饰，限定，使温和
milling ['mɪlɪŋ] n. 铣削
pulling ['pʊlɪŋ] n. 拉销
sawing ['sɔːɪŋ] n. 锯切
filing ['faɪlɪŋ] n. 锉削
generate ['dʒenəreɪt] v. 产生，引起
multidimensional [ˌmʌltɪdaɪ'menʃənl] adj. 多维的
computer-aided process planning 计算机辅助工艺设计
route sheet 工艺图表
spare part 备件
manufacturing resource 制造资源
workpiece center 工件中心
product mix 产品结构
agile manufacturing environment 敏捷制造环境
physical ['fɪzɪk(ə)l] adj. 物理的，身体的，物质的
physical dimension 物理尺寸

differential [ˌdɪfəˈrenʃl] adj. 微分的，差别的，特异的
differential equation 微分方程

 Notes

1. evolve to 发展到，演进到

例句：As the design process is supported by many computer-aided tools, computer-aided process planning (CAPP) has evolved to simplify and improve process planning to achieve more effective use of manufacturing resources.

由于设计过程得到多个计算机辅助工具的支持，计算机辅助工艺设计 (CAPP) 已经发展为简化和改进工艺设计，以实现生产资源的更有效利用。

例句：Why did humans evolve to walk upright?

人类为何进化成直立行走？

2. apply to 应用于，适用于

例句：Generative CAPP uses iterative methods, where simple production plans are applied to automatic CAD/CAM development to refine the initial production plan.

生成式 CAPP 使用迭代方法，将简单的生产计划应用到自动 CAD/CAM 开发中，以完善最初的生产计划。

例句：The convention does not apply to us.

该协定对我们不适用。

3. build on 建立于，以……为基础

例句：Generative CAPP systems are built on a factory's production capabilities and capacities.

生成式 CAPP 系统建立在工厂的产能基础上。

例句：You can easily build on top of this simple example.

你可以轻松地在这个简单示例的基础上进行构建。

4. independent of 不依赖……的，不受……支配的

例句：Each process plan can be defined automatically, independent of past process routes.

每个工艺计划可以自动定义，与过去的工艺路线无关。

例句：It was important to me to be financially independent of my parents.

在经济上不依赖父母，这对我很重要。

5. associated with 与……相关，和……来往

例句：The combination of the feature extensions of the process plan and the parametric data associated with them became part of the data.

工艺设计的特征扩展和相关参数的数据组合成为产品数据管理 (PDM) 系统的数据

的一部分。

例句：We associate with all sorts of people.

我们同各种各样的人交往。

6. be based on 基于，以……为根据

例句：Ultimately, the parameters for this system were based on expenditures, time, physical dimensions, and availability.

最终，该系统的参数基于支出、时间、物理尺寸和可用性。

例句：The book is based on personal experience.

本书是根据个人经历写成的。

 Exercises

Ⅰ. Write True or False beside the following statements about the text.

1. _____ CAPP is the link between CAD and CAM in that it provides for the planning of the process to be used in producing a designed part.

2. _____ The resulting operation sequence is documented on a form typically referred to as a "route sheet" (also called a process sheet) containing a list of the production operations and associated machine tools for a work part or assembly.

3. _____ Generative CAPP systems are not built on a factory's production capabilities and capacities.

4. _____ As improvements are made to production efficiencies, the improvements are not incorporated into the current product mix.

5. _____ Execution planning can be dynamic and accommodate changing conditions.

Ⅱ. Answer the following questions in English according to the text.

1. What does CAPP refer to? Which is the key component of the CAPP system for the agile manufacturing environment?

2. What are generative CAPP systems based on?

3. In order to achieve the generative CAPP system, what components were built to meet the capabilities?

Ⅲ. Read the text again and fill in the blanks in the following sentences orally.

1. Computer-aided process planning (CAPP) is the use of _____ _____ to aid in the _____ _____ of a part or product in manufacturing.

2. CAPP is the link between _____ and _____ in that it provides for the planning of the process to be used in producing a _____ _____.

3. The resulting operation sequence is documented on a form typically referred to as a "_____ _____" (also known as _____ _____) containing a list of the

production operations and associated machine tools for a work part or assembly.

4. Generative CAPP uses _____ _____, where simple production plans are applied to automatic CAD/CAM development to refine the initial production plan.

5. The scheduling is continuously updated based on the real-time conditions of the enterprise. Ultimately, the parameters for this system were based on _____, time, _____ _____, and _____.

Ⅳ. Translation.

Create a manufacturing execution system (MES). The MES's major component is an expert/artificial intelligent system that matches the engineering feature objects from the product data management (PDM) system with the tooling, personnel, material, transportation needs, etc. needed to manufacture them in the enterprise resource planning (ERP) system. Once the above physical components are identified, the process items can be scheduled.

Lesson 19 Automatic Detection of Weld Defects

X-ray imaging is a widely used technique for the inspection of industrial pieces and for the medical diagnosis of human diseases. Traditional radiographic testing (RT) with the film is an expensive and time-consuming technique. Therefore, digital radioscopy that permits real-time inspection has been developed and applied. Taking welding defect detection in the manufacturing industry as an example, now experienced workers are required to evaluate the moving weld seam based on the video displayed on the monitor. However, the manual interpretation process can be subjective and inconsistent, and easily cause fatigue.

Therefore, it is imperative to develop an automatic computer-aided system to increase the objectivity, consistency, and efficiency of defect inspection. Now, the automatic real-time digital radiographic detection system is developed to solve the above problems. This system consists of[①] two aspects: identifying the defects in the weld and classifying different types of welding defects.

The detection system consists of a conversion part, a processing part, and a serial communication part, as shown in Fig. 19-1. The conversion part consists of an X-ray source, a transmission vehicle, an

Margin Note

X-ray imaging
X射线成像

digital radioscopy
数字放射镜

weld seam
焊缝

digital radiographic detection system
数字射线检测系统

intensifier, and a CCD (charge-coupled device) camera. The function of this part is to make the conversion from X-rays to visible a light. Firstly, X-ray is transferred into visual light through a light intensifier. Then the CCD camera transfers the light signal into an electric signal and sends it to the processing part. The processing part consists of a monitor, an image grabber, a computer, and a screen. In this part, the electric signal is sampled and transformed into② a digital signal by an image grabber, and at the same time, it is also displayed on the monitor. The digital image is sent into the computer and will be detected using the defect detection algorithm based on fuzzy recognition theory. The result will be displayed on the screen in real-time and stored in the computer for future checks or tests. The serial communication part consists of a single-chip computer (SCM), a rotary coder, an optical isolator module, a transmission device, etc. The function of this part is to obtain and transmit the information of position. The system transforms the displaced signal into the pulse signal utilizing the rotary coder and attains displacement by computing the number of the pulse. Then the displacement signal is transmitted to the computer for the defect's location through serial communication.

intensifier
增强器
charge
coupled device
电荷耦合器件
image grabber
图像采集器
defect
detection algorithm
缺陷检测算法
fuzzy
recognition theory
模糊识别理论

single-chip computer
单片机
rotary coder
旋转编码器
displaced signal
位移信号
pulse signal
脉冲信号
serial
communication
串行通信

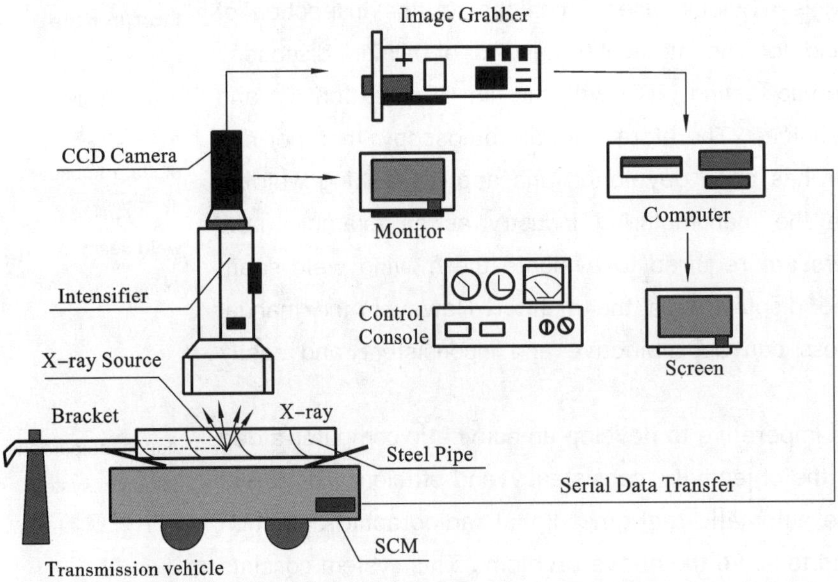

Fig. 19-1 The Component Structure of the Detection System

The defects detection includes two main procedures: image preprocessing and realization of the algorithm. Usually, an automatic welding process yields a relatively uniform weld. Since only the items

image
preprocessing
图像预处理

within a weld are of interest[③] for image processing, this system extracted the weld area free of background. This system has prior information about brightness distribution in the points of object and background. There are three main areas in the weld image: the base metal area, the weld area, and the lead plate area. The weld seam is oriented along the horizontal direction.

After welds are successfully extracted, it is then identified the defect areas, as shown in Fig. 19-2. X-ray imaging is inherently noisy because of the quantum nature of radiation. There may be only a few photons per pixel per exposure time. The large defect can be easily detected by many methods. But small defects encounter the difficulty of differentiation between true defect pixels and noisy impulses. In order to achieve fast and precise detection, a region-based approach that imitates visual inspection is used. For visual inspection, contrast and variance are important because the human visual system is very sensitive to the two parameters. The contrast is usually given by the grey level difference between the object and its neighborhoods. This is the principle of the fuzzy algorithm applied to identify defects in the weld seam. It can give very high confidence about the defect, but rough information on the type.

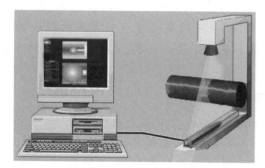

Fig. 19-2 Real-time X-ray Imaging Detection System

X-ray detection is widely used in industry. It is often used to inspect the internal structure of products, such as precision castings, integrated circuits, electronic components, LED components, BGA, circuit boards, and IC chip products, for void content of internal common bubbles, foreign body crack, short circuits, broken circuits, lack of[④] weld, less weld or missing weld at joints, as well as the gap of welding joints in surface mount technology (SMT).

brightness distribution
亮度分布
base metal area
母材区
lead plate area
导板区

true defect pixel
真实缺陷像素

fuzzy algorithm
模糊算法

precision casting
精密铸件
integrated circuit
集成电路
circuit board
电路板
void content
空洞率
short circuit
电路短路

Tips

1. X-ray X 射线

X 射线,是一种频率极高、波长极短、能量很大的电磁波。X 射线在频率和能量方面仅次于伽马射线。X 射线具有穿透性,但人体组织有密度和厚度的差异,当 X 射线透过人体不同组织时,其被吸收的程度不同,经过显像处理后即可得到不同的影像。

2. RT:Radiographic Testing 射线检验

射线检验是应用较早的材料检测方法之一。当强度均匀的射线束透射物体时,如果物体局部区域存在缺陷或结构存在差异,则会引起物体对射线的衰减,使得不同部位的透射强度不同。这样,采用一定的检测器(例如,射线照相中采用胶片)检测透射强度,就可以判断物体内部的缺陷和物质分布等,从而完成对被检测对象的检验。

3. CCD:Charge Coupled Device 电荷耦合器件

电荷耦合器件(CCD),是一种用电荷量表示信号大小、用耦合方式传输信号的探测元件,具有自扫描、感受波谱范围宽、畸变小、体积小、重量轻、系统噪声低、功耗少、寿命长、可靠性高等一系列优点,并可做成集成度非常高的组合件。CCD 广泛应用在数码摄影、天文学领域,尤其是光学遥测技术、光学与频谱望远镜和高速摄影技术中。

4. SCM:Single-chip Computer 单片机

单片机又称单片微控制器,它是指把一个计算机系统集成到一个芯片上。由于单片机的集成度高,因此单片机具有体积小、功耗少、控制功能强、扩展灵活、微型化和使用方便等优点,被广泛应用于智能仪器仪表制造、工业控制、家用智能电器制造、网络通信设备使用和医疗卫生行业。

5. BGA:Ball Grid Array 球栅阵列封装

球栅阵列封装技术为应用在集成电路上的一种表面黏着封装技术,此技术常用来永久固定如微处理器之类的装置。球栅阵列封装是一种将某个表面以格状排列的方式覆满(或部分覆满)引脚的封装法,在运作时即可将电子信号从集成电路上传导至其所在的印刷电路板。在球珊阵列封装下,在封装底部的引脚由锡球所取代,每个原本是一粒小小的锡球固定其上。

6. SMT: Surface Mounting Technology 表面贴装技术

SMT 是在 PCB(printed circuit board,印刷电路板)基础上进行加工的系列工艺流程的简称。SMT 是电子组装行业里最流行的一种技术和工艺,它是一种将无引脚或短引线表面装配元件(简称 SMC/SMD,中文称片状元件)安装在印刷电路板的表面或其他基板的表面上,通过再流焊或浸焊等方法进行焊接装配的电路装连技术。在通常情况下,我们用的电子产品都是由 PCB 加上各种电容器、电阻器等电子元件按电路图设计而成的。

New Words and Phrases

radiographic [ˌreɪdɪəʊˈɡræfɪk] adj. 胶片照相术的,射线照相术的
digital radiographic detection system 数字射线检测系统
radioscopy [ˌreɪdɪˈɒskəpɪ] n. 射线检查法,X 光透视,放射线透视
digital radioscopy 数字放射镜
evaluate [ɪˈvæljueɪt] v. 评价,评估,估值
weld seam 焊缝
subjective [səbˈdʒektɪv] adj. 主观的,个人的,自觉的
inconsistent [ˌɪnkənˈsɪstənt] adj. 不一致的,不协调的,前后矛盾的
fatigue [fəˈtiːɡ] n. 疲乏,厌倦,(金属部件的)疲劳
imperative [ɪmˈperətɪv] adj. 必要的,命令的,强制的
defect [ˈdiːfekt] n. 缺点,缺陷,毛病
defect detection algorithm 缺陷检测算法
conversion [kənˈvɜːʃn] n. 转换,转变,改变,归附
transform [trænsˈfɔːm] v. 转换,使变形,使转化
fuzzy [ˈfʌzi] adj. (图片、声音等)不清楚的,模糊的
fuzzy algorithm 模糊算法
isolator [ˈaɪsleɪtə(r)] n. 隔离器,隔音装置,绝缘体
displaced [dɪsˈpleɪst] adj. 位移的,被取代的,无家可归的
displaced signal 位移信号
pulse signal 脉冲信号
procedure [prəˈsiːdʒə(r)] n. 手续,步骤
yield [jiːld] v. 产生,放弃,让步 n. 产量,收益,利润,红利
inherently [ɪnˈherəntli] adv. 内在地,固有地
quantum [ˈkwɒntəm] n. 量子,量子论,额(特指定额、定量)
void [vɔɪd] n. 空洞,空隙,空白 adj. 不合法的
void content 空洞率

approach [əˈprəʊtʃ] n. 方法,态度,靠近,接近
variance [ˈveərɪəns] n. 变化幅度,分歧,不一致,方差
casting [ˈkɑːstɪŋ] n. 铸件,铸造物,角色分配,演员挑选
precision casting 精密铸件
X-ray imaging X 射线成像
intensifier [ɪnˈtensɪfaɪə(r)] n. 增强器
coupled [ˈkʌp(ə)ld] adj. 耦合的,联结的
charge coupled device 电荷耦合器件
grabber [ˈɡræbə(r)] n. 采集器,掠夺者
image grabber 图像采集器
recognition [ˌrekəɡˈnɪʃn] n. 识别,承认,表彰
fuzzy recognition theory 模糊识别理论
single-chip computer 单片机
rotary coder 旋转编码器
serial communication 串行通信
image preprocessing 图像预处理
brightness distribution 亮度分布
base metal area 母材区
lead [liːd] v. 引领,带路 n. 领先地位
lead plate area 导板区
true defect pixel 真实缺陷像素
integrated circuit 集成电路
circuit board 电路板
short circuit 电路短路

 Notes

1. consist of 由……组成

例句:This system <u>consists of</u> two aspects:identifying the defects in the weld and classifying different types of welding defects.

该系统由两个方面<u>组</u>成:对焊缝缺陷进行识别和对不同类型的焊接缺陷进行分类。

例句:A student's education does not only <u>consist of</u> learning academic subjects.

对学生的教育不仅仅<u>包含</u>学术科目的学习。

2. transform into 转变成

例句:In this part,the electric signal is sampled and <u>transformed into</u> a digital signal by an image grabber.

该部分通过图像采集器对电信号进行采样,并将其<u>转换</u>为数字信号。

例句：These awesome earphones from Takara Tomy can be transformed into a robot!

这款出自多美公司的非常棒的耳机可以变形成机器人！

3. be of interest 关注，吸引

例句：Since only the items within a weld are of interest for image processing, this system extracted the weld area free of background.

由于图像处理的目的只关注焊缝这一项，因此该系统提取无背景的焊缝区域。

例如：With its vivid description of the life of ancient Chinese people, the book will be of interest to a wide range of readers.

因为对中国古代人民生活的生动描写，这本书将会吸引一大批读者。

4. lack of 没有，缺乏，不足，不够

例句：For void content of internal common bubbles, foreign body crack, short circuits, broken circuits, lack of weld, less weld or missing weld at joints, as well as the gap of welding joints in surface mount technology (SMT).

用于表面贴装技术（SMT）中内部常见气泡、异物裂缝、电路短路断路、焊点缺焊少焊漏焊等的检查。

例句：They were exhausted from lack of sleep.

他们因睡眠不足而疲惫不堪。

 Exercises

Ⅰ. Write True or False beside the following statements about the text.

1. _____ X-ray imaging is a widely used technique for the inspection of industrial pieces and the medical diagnosis of human diseases.

2. _____ Therefore, it is not necessary to develop an automatic computer-aided system to increase the objectivity, consistency, and efficiency of defect inspection.

3. _____ The defects detection includes two main procedures: image preprocessing and realization of the algorithm.

4. _____ There are three main areas in the weld image: the base metal area, the operation area, and the lead plate area.

5. _____ To achieve fast and precise detection, a region-based approach that imitates visual inspection is used.

Ⅱ. Answer the following questions in English according to the text.

1. What is automatic real-time digital radiographic detection system used for?
2. How many parts does the detection system consist of? What are they?
3. How does the defect detection system identify the defect areas?
4. What areas could X-ray detection apply to?

Ⅲ. **Read the text again and fill in the blanks in the following sentences orally.**

1. Taking welding defect detection in the _____ _____ as an example, now experienced workers are required to evaluate the moving _____ _____ based on the video displayed on the monitor.

2. Therefore, it is imperative to develop an _____ _____ system to increase the _____, consistency, and _____ of defect inspection.

3. The conversion part consists of an _____ _____, a transmission vehicle, an _____ and a CCD (charge-coupled device) camera.

4. The system transforms the _____ signal into the _____ signal utilizing the _____ _____ and attains displacement by computing the number of the pulse.

5. For visual inspection, _____ and _____ are important because the human visual system is very sensitive to the two parameters. The contrast is usually given by the _____ _____ difference between the object and its neighborhoods.

Ⅳ. **Translation.**

X-ray detection is widely used in industry. It is often used to inspect the internal structure of products, such as precision castings, integrated circuits, electronic components, LED components, BGA, circuit boards, and IC chip products, for void content of internal common bubbles, foreign body crack, short circuits, broken circuits, lack of weld, less weld or missing weld at joints, as well as the gap of welding joints in surface mount technology (SMT).

Lesson 20 Intelligent Manufacturing Control MES Software

Intelligent manufacturing control MES (manufacturing execution system) software is a control system deployed on a computer and used in automatic production lines. It monitors the operation of machine tools, robots, measuring instruments, and other equipment on the production line and provides a convenient visual interface to display the detected data. At the same time, the intelligent production line MES system can complete the upload and download of data, report the data (work report, status, action, tool, etc.), and issue production tasks and commands (CNC cut-in and cut-out, control instructions, machining tasks) to the equipment.

Margin Note

automatic production line
自动生产线

production line
生产线

Human machine interface (HMI) constitution is shown in Fig. 20-1. ①Menu bar: the menu bar, including the File Selection button, Edit button, View button, Parameter button, and Help button. Among them, "Edit" can change the background color of the software and switch between Chinese and English as needed, and "Help" can view the software version. ②Machine tool equipment status: real-time current machine tool operating status, including offline, idle, running, and alarm status. ③Equipment status area: real-time current online status of machine tools, PLC(programmable logic controller), and robots. ④Tag bar area: display the tags of each function page. ⑤Function display and settings area: display the main content under the current function tab, including related data display and settings. ⑥System time bar: display the current system time, the cumulative running time of the system, and the alarm information. ⑦PLC reconnection: when the PLC shows that the status is offline, clicking the PLC Reconnect button, MES software will attempt to① connect the PLC. ⑧Production line starts: click the Production Line Start button, when the production line starts, you can issue orders for processing; the Production Line Start button is valid under the user login, if the user does not log in, the production line start will give a login prompt. ⑨Production line stop: click the Production Line Stop button, and the production line stops cannot issue orders. ⑩Line reset: click the Line Reset button, and the machine tool executes the HOME program, and reset the device.

programmable
logic controller
可编程控制器

alarm information
报警信息

login prompt
登录提示

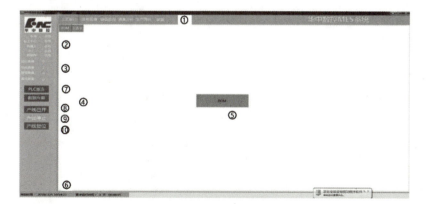

Fig. 20-1　The Page of the Human Machine Interface of MES Software

Intelligent manufacturing control MES software mainly has the following functions:

(1)Monitoring equipment. Intelligent production line MES system is mainly used to monitor the operation of the production line equipment and production and release tasks, the main detection objects are machine tools, RFID, industrial robots, silos, and measuring instruments.

(2) Scheduling management. The main production schedule has the function of order execution. It is used to generate, deliver, cancel, and delete. And it is also used to select the execution operation of the workpiece order, including feeding, blanking, refueling, automatic processing, and other operations.

(3) Equipment monitoring. The machine tool page displays the machine tool information, such as connection status, IP, port, system version, and the related parameter information of the machine tool system, including machine tool system information, machine tool operation, tool information, etc. It is used to display the robot's axis position information, state information, working mode, whether it is at home point, etc. It is also used to display silo information, as shown in Fig. 20-2, and control silo five-color lights, including bin status monitoring, processing program monitoring, bin inventory, material level initialization, and five-color lamp control. On the Camera Configuration and Display page, set camera parameters and display camera content. The monitor page includes login settings, preview, action information, event callback information, and alarms, as shown in Fig. 20-3.

silo information
料仓信息

bin status monitoring
料仓状态监控

bin inventory
料仓盘点

material level initialization
料位初始化

login setting
登录设置

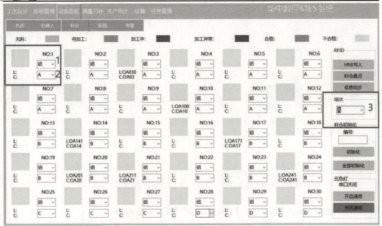

Fig. 20-2 Digital Silo Information

Fig. 20-3　Configure the Camera Parameters of the Monitoring Device

 New Words and Phrases

deploy [dɪˈplɔɪ] v. 部署,有效利用
automatic production line 自动生产线
interface [ˈɪntəfeɪs] n. 界面,接口
issue [ˈɪʃuː] v. 下发,宣布,公布
idle [ˈaɪdl] adj. 空闲的,闲置的
prompt [prɒmpt] n. 提示,提示符
login prompt 登录提示
reset [ˌriːˈset] v. 复位,重置
refuel [ˌriːˈfjuːəl] v. 换料,加燃料,加油
silo [ˈsaɪləʊ] n. 料仓,筒仓
silo information 料仓信息
bin [bɪn] n. 仓,箱子
bin inventory 料仓盘点
bin status monitoring 料仓状态监控
initialization [ɪˌnɪʃəlaɪˈzeɪʃn] n. 初始化
material level initialization 料位初始化
configuration [kənˌfɪɡəˈreɪʃn] n. 配置,结构,构造
lathe [leɪð] n. 车床,机床
programmable logic controller 可编程控制器

alarm information 报警信息
login setting 登录设置

Notes

1. attempt to 尝试，努力

例句：When the PLC shows that the status is offline, clicking the PLC Reconnect button, MES software will <u>attempt to</u> connect the PLC.

当 PLC 显示为离线状态时，单击"重新连接"按钮，MES 软件将<u>尝试</u>连接 PLC。

例句：I will <u>attempt to</u> answer all your questions.

我将<u>努力</u>回答你的全部问题。

Exercises

Ⅰ. Write True or False beside the following statements about the text.

1. _____ Intelligent manufacturing control MES (manufacturing execution system) software is a control system deployed on a cell phone.

2. _____ System time bar: display the current system time, the cumulative running time of the system, and the alarm information.

3. _____ If the user logs in, the production line start will give a login prompt.

4. _____ The monitor page includes login settings, preview, action information, event callback information, and alarms.

5. _____ To import data in batches, select AutoCAD file.

Ⅱ. Answer the following questions in English according to the text.

1. What is intelligent manufacturing control MES (manufacturing execution system) software?

2. What does the machine tool page display?

3. What does the monitor page include?

Ⅲ. Read the text again and fill in the blanks in the following sentences orally.

1. "Edit" can change the _____ color of the software and switch between Chinese and English as needed.

2. Equipment status area: real-time current _____ status of machine tools, PLC, and robots.

3. On the Camera Configuration and Display page, _____ camera parameters and _____ camera content.

Ⅳ. Translation.

Intelligent manufacturing control MES (manufacturing execution system) software is a control system deployed on a computer and used in automatic production lines. It monitors the operation of machine tools, robots, measuring instruments, and other equipment on the production line and provides a convenient visual interface to display the detected data. At the same time, the intelligent production line MES system can complete the upload and download of data, report the data (work report, status, action, tool, etc.), and issue production tasks and commands (CNC cut-in and cut-out control instructions, machining tasks) to the equipment.

Reading Material 13 Product Data Management

Product data management (PDM) is the technology and associated software systems that support the management of both engineering data and process information during the product development phase and beyond. As shown in Fig. R13-1, the data collected for product data management is processed for deployment to be used in other strategic systems throughout the company. The other systems make up product lifecycle management. PDM aims at providing product design teams with the right data and information at the right time for making proper design decisions.

Margin Note

product life cycle management
产品生命周期管理
distributed network
分布式网络

Fig. R13-1 The Product Data Management Structure

Product data management (PDM) is a software framework (or data platform) developed based on distributed network, master-slave structure, graphical user interface, and database component management technology. PDM comprehensively manages personnel,

master-slave structure
主从结构
graphical user interface
图形化用户接口

tools, equipment resources, product data, and data generation processes in concurrent engineering.

In the past, enterprises use CAD, CAM, and other technologies in their design and production process, they are all self-contained and lack effective information sharing and utilization between each other, forming the so-called "information island". In this case, many enterprises have realized that the orderly management of information will become the key factor to keeping ahead in the future competition. Product data management (PDM) is a new management idea and technology under this background. PDM can be defined as a technology that realizes the integrated management of product-related data, processes, and resources based on software technology and centering on products. This is also because the PDM system is different from other information management systems, such as enterprise management information system (MIS), manufacturing resource planning (MRP), project management (PM) system, and enterprise resource planning (ERP) system.

concurrent engineering 并行工程

The product data management system has the following main functions:

(1) Electronic database and document management.

(2) Product structure and configuration management.

(3) Life cycle (workflow) management and production cycle management.

configuration management 配置管理

(4) Development-interface integration.

PDM systems capture and manage product information, ensuring that information is delivered to users throughout the product life cycle in the correct context. PDM systems provide the visibility necessary for managing and presenting a complete bill of materials (BOM). It facilitates the alignment and synchronization of all sources of BOM data, as well as all life cycle phases, including the as-designed, as-planned, as-built, and as-maintained states, as shown in Fig. R13-2.

PDM covers a wide range of fields, applicable product areas include:

(1) Manufacturing—automobiles, airplanes, ships, computers, home appliances, mobile phones, etc.

(2) Engineering projects—buildings, bridges, highways.

(3) Factories—steel works, oil refineries, food processing plants, pharmaceutical plants, offshore platforms, etc.

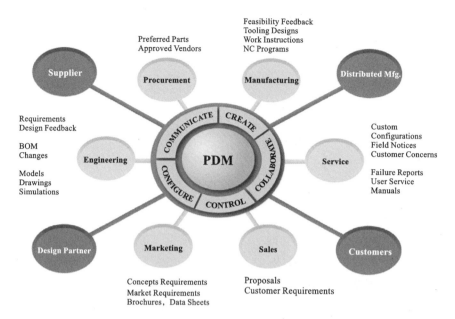

Fig. R13-2　The Diagram of Product Data Management Structure

（4）Infrastructure—airports, seaports, rail operation systems, logistics warehousing.

（5）Public utilities—power generation/power setup, wireless communications, water/coal/gas supply, cable TV network.

（6）Finance—banking, securities trading, and other industries.

pharmaceutical plant
制药厂
offshore platform
海洋平台

Within PDM the focus is on managing and tracking the creation, change, and archive of all information related to a product. The information being stored and managed (on one or more file servers) will include engineering data such as computer-aided design (CAD) models, drawings, and their associated documents.

file server
文件服务器

Typical information managed within the PDM module includes:

(1) Brand name.

(2) Part number.

(3) Part description.

(4) Supplier/vendor.

(5) Vendor part number and description.

(6) Unit of measure.

(7) Cost/price.

(8) Schematic or CAD drawing.

(9) Material data-sheets.

(10) Advantages.

(11) Track and manage all changes to product-related data.

(12) Spend less time organizing and tracking design data.
(13) Improve productivity through the reuse of product design data.
(14) Enhance collaboration.
(15) Help using visual management.

New Words and Phrases

deployment [dɪˈplɔɪmənt] n. 部署，调集
self-contained 自成体系的，独立的，自立的
alignment [əˈlaɪnmənt] n. 对齐，排成直线
approved vendor 认可供应商
feasibility feedback 可行性反馈
infrastructure [ˈɪnfrəstrʌktʃə(r)] n. 基础设施
product life cycle management 产品生命周期管理
distributed network 分布式网络
master-slave structure 主从结构
graphical user interface 图形化用户接口
concurrent engineering 并行工程
configuration management 配置管理
pharmaceutical plant 制药厂
offshore platform 海洋平台
file server 文件服务器

Reading Material 14   Manufacturing Resource Planning

Manufacturing resource planning (MRP) is defined as a method for the effective planning of all resources of a manufacturing company, as shown in Fig. R14-1. Ideally, it addresses operational planning in units, and financial planning, and has a simulation capability to answer "what-if" questions.

This is not exclusively a software function, but the management of people skills, requiring a dedication to database accuracy, and sufficient computer resources. It is a total company management concept for using

Margin Note

vendor order
供应商订单

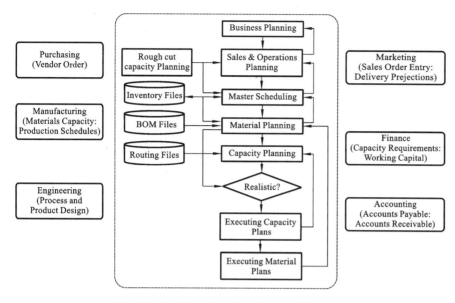

Fig. R14-1 Manufacturing Resource Planning

human and company resources more productively.

Almost every MRP system is modular in construction. Characteristic basic modules in an MRP system are:

(1) Master production schedule (MPS).
(2) Item master data (technical data).
(3) Bill of materials (BOM) (technical data).
(4) Production resources data (manufacturing technical data).
(5) Inventories and orders (inventory control).
(6) Purchasing management.
(7) Material requirements planning (MRP).
(8) Shop floor control (SFC).
(9) Capacity planning or capacity requirements planning (CRP).
(10) Standard costing (cost control) and frequently also Actual or FIFO costing, and Weighted Average costing.
(11) Cost reporting / management (cost control).

Together with auxiliary systems such as:

(1) Business planning.
(2) Lot traceability.
(3) Contract management.
(4) Tool management.
(5) Engineering change control.
(6) Configuration management.
(7) Shop floor data collection.

rough cut capacity planning
粗略缩减容量规划
inventory file
库存文件
BOM file
物料清单文件
routing file
工艺路线文件
sales & operations planning
销售和运营计划
master scheduling
主调度
executing capacity plan
执行产能计划
executing material plan
执行物料计划
sales order entry
销售订单输入
delivery projection
交付预测
working capital
营运资金
accounts payable
应付账款
accounts receivable
应收账款
capacity planning
容量规划
weighted average costing
加权平均成本法

configuration management
配置管理

(8) Sales analysis and forecasting.

(9) Finite capacity scheduling (FCS). finite capacity scheduling 有限容量调度

And related systems such as:

(1) General ledger.

(2) Accounts payable (purchase ledger).

(3) Accounts receivable (sales ledger).

(4) Sales order management.

(5) Distribution resource planning (DRP).

(6) Automated warehouse management.

(7) Project management.

(8) Technical records.

(9) Estimating.

(10) Computer-aided design/computer-aided manufacturing (CAD/CAM).

(11) CAPP.

The MRP system integrates these modules so that they use common data and freely exchange information, in a model of how a manufacturing enterprise should and can operate. The MRP approach is therefore very different from the "point solution" approach, where individual systems are deployed to help a company plan, control or manage a specific activity. MRP is by definition fully integrated or at least fully interfaced.

Like today's ERP systems, MRP was designed to tell us a lot of information by way of a centralized database. However, the hardware, software, and relational database technology of the 1980s was not advanced enough to provide the speed and capacity to run these systems in real time, and the cost of these systems was prohibitive for most businesses. Nonetheless, the vision had been established, and shifts in the underlying business processes along with rapid advances in technology led to the more affordable enterprise and application integration systems that big businesses and many medium and smaller businesses use today. relational database technology 关系数据库技术

MRP is concerned with the coordination of the entire manufacturing production, including materials, finance, and human resources. The goal of MRP is to provide consistent data to all members in the manufacturing process as the product moves through the production line.

MRP systems can provide:

(1) Better control of inventories.

(2) Improved scheduling.
(3) Productive relationships with suppliers.

For design / engineering：
(1) Improved design control.
(2) Better quality and quality control.

For financial and costing：
(1) Reduced working capital for inventory.
(2) Improved cash flow through quicker deliveries.
(3) Accurate inventory records.

New Words and Phrases

rough cut capacity planning 粗略缩减容量规划
inventory file 库存文件
routing file 工艺路线文件
sales & operations planning 销售和运营计划
master scheduling 主调度
executing capacity plan 执行产能计划
executing material plan 执行物料计划
marketing [ˈmɑːkɪtɪŋ] n. 促销，营销
sales order entry 销售订单输入
delivery projection 交付预测
finance [ˈfaɪnæns] n. 财务，资金，财政，金融
capacity requirement 能力要求
working capital 营运资金
accounts payable 应付账款
accounts receivable 应收账款
dedication [ˌdedɪˈkeɪʃn] n. 投入，奉献精神
accuracy [ˈækjərəsi] n. 精确程度，准确性
integrate [ˈɪntɪɡreɪt] v. 集成，整合，(使)合并
prohibitive [prəˈhɪbətɪv] adj. 令人望而却步的，禁止的，贵得买不起的
nonetheless [ˌnʌnðəˈles] adv. 尽管如此，不过

Reading Material 15 Quality Control

Quality control refers to the activities of technical and management measures taken to make products or services meet quality requirements. The goal of quality control is to ensure that the quality of products or services can meet the requirements and standards.

1. Tools and Techniques for Quality Control

(1) Inspection. Inspection includes activities such as measuring, examining, and testing is undertaken to determine whether results conform to requirements. Inspections may be conducted at any level (e.g., the results of a single activity may be inspected, or the final product of the project may be inspected). Inspections are variously called reviews, product reviews, audits, and walkthroughs: in some application areas, these terms have narrow and specific meanings.

(2) Control charts. Control charts are a graphic display of the results of process over time. They are used to determine if the process is "in control" (e.g., are differences in the results created by random variations, or are unusual events occurring whose causes must be identified and corrected). When a process is in control, the process should not be adjusted. The process may be changed to provide improvements, but it should not be adjusted when it is in control. Control charts may be used to monitor any type of output variable. Although used most frequently to track repetitive activities, such as manufactured lots, control charts can also be used to monitor cost and schedule variances, volume and frequency of scope changes, errors in project documents, or other management results to help determine if the project management process is in control. Fig. R15-1 is a control chart of project schedule performance.

(3) Pareto diagrams. A Pareto diagram is a histogram, ordered by frequency of occurrence, which shows how many results were generated by type or category of identified cause (as shown in Fig. R15-2). Rank order is used to guide corrective action—the project team should take action to fix the problems that are causing the greatest number of defects first. Pareto diagrams are conceptually related to Pareto's law, which holds that a relatively small number of causes will typically

Margin Note

audit
审计

walkthrough
巡回检查

control chart
控制图

random variation
随机变化

Pareto diagram
帕累托图

Pareto's law
帕累托法则

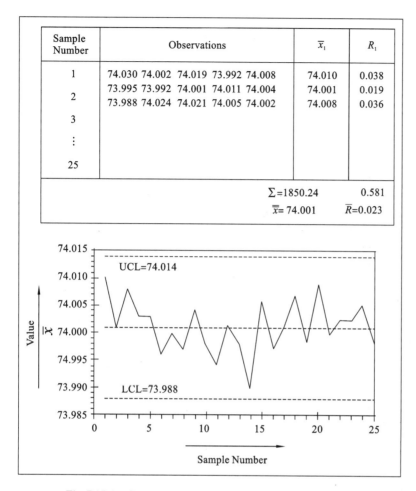

Fig. R15-1 Control Chart of Project Schedule Performance

produce a large majority of the problems or defects. This is commonly referred to as the 80/20 principle, where 80 percent of the problems are due to 20 percent of the causes.

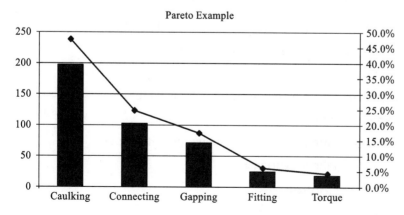

Fig. R15-2 Pareto Diagram

(4) Statistical sampling. Statistical sampling involves choosing part of a population of interest for inspection (e.g., selecting ten engineering drawings at random from a list of seventy-five). Appropriate sampling can often reduce the cost of quality control. There is a substantial body of knowledge on statistical sampling; in some application areas, the project management team must be familiar with a variety of sampling techniques.

statistical sampling
统计抽样

(5) Flowcharting. Flowcharting is used in quality control to help analyze how problems occur, as shown in Fig. R15-3.

flowcharting
流程图

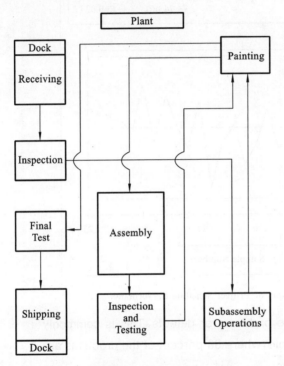

Fig. R15-3　Flow Chart

(6) Trend analysis. Trend analysis involves using mathematical techniques to forecast future outcomes based on historical results. Trend analysis is often used to monitor:

①Technical performance — how many errors or defects have been identified, how many remain uncorrected.

technical performance
技术绩效

②Cost and schedule performance — how many activities per period were completed with significant variances.

2. Outputs from Quality Control

(1) Quality improvement.

(2) Acceptance decisions. The items inspected will be either

rework
返工

accepted or rejected. Rejected items may require rework.

(3) Rework. Rework is an action taken to bring a defective or nonconforming item into compliance with requirements or specifications. Rework, especially unanticipated rework is a frequent cause of project overruns in most application areas. The project team should make every reasonable effort to minimize rework.

project overrun
项目超支

(4) Completed checklists. When checklists are used, the completed checklists should become part of the project's records.

checklist
检查表

(5) Process adjustments. Process adjustments involve immediate corrective or preventive action as a result of quality control measurements. In some cases, the process adjustment may need to be handled according to procedures for integrated change control.

New Words and Phrases

conform [kən'fɔːm] vi. 符合，遵照，适应环境
review [rɪ'vjuː] vt, n. 审查
audit ['ɔːdɪt] vi, n. 审计，查账
walkthrough [wɔkθru] n. 巡回检查
schedule ['ʃedjuːl] n. 计划(表)，进度表，时间表
chart [tʃɑːt] n. 图表
graphic display 图形显示
random ['rændəm] adj. 随机的，任意的，胡乱的
variable ['veərɪəb(ə)l] n. 变量，变数
repetitive [rɪ'petətɪv] adj. 重复的
histogram ['hɪstəɡræm] n. 直方图，柱状图
defective [dɪ'fektɪv] adj. 有问题的，有缺陷的
overrun ['əʊvərʌn] n. 超出的成本(费用)，超出的时间
category ['kætɪɡ(ə)rɪ] n. 种类，分类，类别
statistical [stə'tɪstɪk(ə)l] adj. 统计的，统计学的
substantial [səb'stænʃl] adj. 大量的，价值巨大的，重大的
control chart 控制图
random variation 随机变化
Pareto diagram 帕累托图
Pareto's law 帕累托法则

statistical sampling 统计抽样
flowchart ['fləʊtʃɑːt] n. 流程图
technical performance 技术绩效
acceptance decision 验收决定
rework [ˌriːˈwɜːk] v. 返工
project overrun 项目超支
checklist ['tʃeklɪst] n. 检查表

Reading Material 16　Enterprise Resource Planning

Enterprise resource planning（ERP）is the integrated management of main business processes, often in real-time and mediated by software and technology. ERP is usually referred to as a category of business management software—typically a suite of integrated applications—that an organization can use to collect, store, manage, and interpret data from many business activities. ERP systems can be locally based or cloud-based. Cloud-based applications have grown in recent years due to information being readily available from any location with internet access.

ERP provides an integrated and continuously updated view of core business processes using common databases maintained by a database management system. The applications that make up the system share data across various departments（manufacturing, purchasing, sales, accounting, etc.）that provide the data. ERP facilitates information flow between all business functions and manages connections to outside stakeholders.

The ERP system integrates varied organizational systems and facilitates error-free transactions and production, thereby enhancing the organization's efficiency. However, developing an ERP system differs from traditional system development. ERP systems run on a variety of computer hardware and network configurations, typically using a database as an information repository.

An enterprise resource planning system is the glue of the different computer systems in a large organization. With ERP software, each department still has its system, but all of the systems can be accessed

Margin Note

information
repository
信息存储库

through one application with one interface.

ERP applications also allow the different departments to communicate and share information more easily with the rest of the company. It collects information about the activity and state of different divisions, making this information available to other parts, where it can be used productively.

ERP applications can help a corporation become more self-aware by linking information about production, finance, distribution, and human resources together. Because it connects different technologies used by each part of a business, an ERP application can eliminate costly duplicate and incompatible technology. The process often integrates accounts payable, stock control systems, order-monitoring systems, and customer databases into one system.

The most fundamental advantage of ERP is that the integration of a myriad of business processes saves time and expense. Management can make decisions faster and with fewer errors. Data becomes visible across the organization. Tasks that benefit from this integration include:

(1) Sales forecasting, which allows inventory optimization.

(2) Chronological history of every transaction through relevant data compilation in every area of operation.

(3) Order tracking, from acceptance through fulfillment.

(4) Revenue tracking, from invoice through cash receipt.

(5) Matching purchase orders (what was ordered), inventory receipts (what arrived), and costing (what the vendor invoiced).

ERP systems centralize business data, which:

(1) Eliminates the need to synchronize changes between multiple systems—consolidation of finance, marketing, sales, human resource, and manufacturing applications.

(2) Brings legitimacy and transparency to each bit of statistical data.

(3) Facilitates standard product naming/coding.

(4) Provides a comprehensive enterprise view (no "islands of information"), making real-time information available to management anywhere, anytime to make proper decisions.

(5) Protects sensitive data by consolidating multiple security systems into a single structure.

ERP provides increased opportunities for collaboration. Data takes many forms in the modern enterprise, including documents, files, forms,

inventory optimization
库存优化
data compilation
数据汇编
order tracking
订单跟踪

audio and video, and emails.

ERP is a centralized system that provides tight integration with all major enterprise functions, such as HR, planning, procurement, sales, customer relations, finance, or analytics, as well as other connected application functions. In that sense, ERP could be described as a "Centralized Integrated Enterprise System (CIES)".

New Words and Phrases

stakeholder [ˈsteɪkhəʊldə(r)] n. 利益相关者,参与人,参与方
glue [gluː] n. 黏合剂,胶水
access [ˈækses] v. 访问,存取(计算机文件),到达,进入,使用
eliminate [ɪˈlɪmɪneɪt] v. 消除,排除,清除
duplicate [ˈdjuːplɪkeɪt, ˈdjuːplɪkət] adj. 重复的,完全一样的,复制的,副本的
incompatible [ˌɪnkəmˈpætəbl] adj. 不兼容的, 不相容的
fundamental [ˌfʌndəˈmentl] adj. 根本的,基本的
myriad [ˈmɪriəd] n. 大量,无数
compilation [ˌkɒmpɪˈleɪʃn] n. 汇编,编写
revenue [ˈrevənjuː] n. 收入,收益,财政收入,税收收入
synchronize [ˈsɪŋkrənaɪz] v. 使同步,在时间上一致,同步进行
consolidation [kənˌsɒlɪˈdeɪʃn] n. 整合,巩固
legitimacy [lɪˈdʒɪtɪməsi] n. 合法性,正统
sensitive [ˈsensətɪv] adj. 敏感的,体贴的,体恤的
procurement [prəˈkjuːmənt] n. 采购,(尤指为政府或机构)购买
information repository 信息存储库
inventory optimization 库存优化
data compilation 数据汇编
order tracking 订单跟踪

Unit Ⅴ   Advanced Manufacturing Process

 Introduction to the Unit 单元导言

先进制造技术是集机械、电子、信息、材料、能源和管理等先进技术于一体的高新技术,已经形成了完整的体系结构。先进制造技术在国民经济中具有重要的支撑作用,承担着为各行各业提供设备、工具和检测仪器的重要任务。本单元将重点讨论有关先进制造技术方面的内容。由于篇幅有限,本单元主要讨论以下内容:微细加工、激光加工、电火花加工、增材制造、数控加工中心手工编程实训、3D 打印技术、纳米加工技术、高速加工、柔性制造系统等。本单元的目的是在学生现有的专业技术知识基础上,结合英语来使学生学习部分新技术、新知识,初步掌握加工技术的英语表达,从而为英语融合专业提供平台。

 Lesson 21   Micromachining

Micromachining is a method of making extremely tiny mechanical or electrical components. This process typically uses wafers of silicon, but will occasionally use plastic or ceramic material as well. Bulk micromachining starts with a solid piece and removes material until it reaches its final shape, as opposed to① surface micromachining, which builds a piece layer by layer. The most common method for performing bulk micromachining is via selective masking and wet chemical solvents, namely chemical etching. The newer alternative to this method is dry etching using a plasma or laser system to remove unwanted material. This is generally more accurate than wet etching, but it is also more expensive.

Machine complex systems of gears and levers are also typical examples of the application of micromachining. The fabrication of these devices is usually done by two techniques: surface micromachining (as shown in Fig. 21-1) and bulk micromachining (as shown in Fig. 21-2).

The vast majority of bulk micromachining uses silicon. The material

Margin Note

surface
micromachining
表面微细加工

chemical etching
化学腐蚀法

wet etching
湿蚀刻法

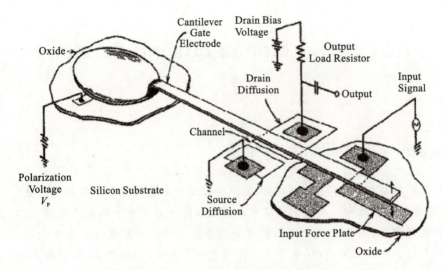

Fig. 21-1 Surface Micromachining of Chips

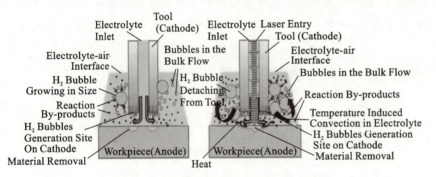

Fig. 21-2 Electrochemical Bulk Micromachining

is extremely cheap since it makes up nearly a quarter of the earth's crust. In addition, it possesses a very fine crystalline structure that can break down into[②] layers thinner than human hair. This allows the material to work on a microscopic level just as well as on a macroscopic.

The most common method for bulk micromachining is called wet chemical etching (as shown in Fig. 21-3). First, the workpiece is covered in a material that will protect it from a selected solvent. The protective mask is then selectively removed to expose the areas of the piece that are coming off. The workpiece is exposed to a solvent, which will then dissolve any unprotected areas and leave the rest intact. Afterward, the remaining masking material is removed.

The latest method for bulk micromachining is called dry etching (as shown in Fig. 21-4). It uses a high-precision device, often a laser, to vaporize unwanted material. Compared with the wet process, this

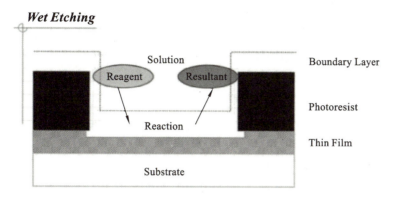

Fig. 21-3 Wet Etching Processes

process has fewer steps and no potentially hazardous solvents.

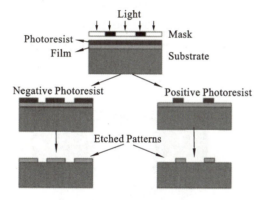

Fig. 21-4 Dry Etching Processes

Surface micromachining is a fabrication process used to develop integrated circuits and sensors of various kinds. Using surface micromachining techniques allows applications of up to nearly 100 finely applied layers of circuit patterns on one chip. This allows many more functions and electronics to be incorporated into[③] each chip for use in motion sensors, vehicle airbag accelerometers, or for use in navigation system gyroscopes. Surface micromachining uses selected materials and both wet and dry etching processes to form the circuitry layers.

The surface micromachining process uses either crystal silicon chip substrates as a foundation upon which to build layers, or can be started on cheaper glass or plastic substrates. Usually, the first layer is of silicon oxide, an insulator, which is etched to a desired thickness. Over this layer, a photosensitive film layer is applied, and ultraviolet (UV) light is applied through the circuit pattern overlay. Next, this wafer is developed, rinsed, and baked for the following etching process. This process is repeated multiple times to apply more layers, with careful

integrated circuit
集成电路

motion sensor
运动传感器
vehicle airbag accelerometer
车辆安全气囊加速计
navigation system gyroscope
导航系统陀螺仪

plastic substrate
塑料基板

monitoring and precise etching techniques applied to each layer, to produce the final layered chip design.

silicon oxide
二氧化硅

The actual surface micromachining process of etching is done by one or a combination of several machining processes. Wet etching is done using hydrofluoric acids to etch out circuit designs on layers, cutting through unprotected insulating materials; un-etched areas of that layer are then electrolyzed to isolate the layer from the next one applied. Dry etching can be done alone, or in combination with[④] chemical etching, using an ionized gas to bombard the areas to be etched. Manufacturers use dry plasma etching (as shown in Fig. 21-5 and Fig. 21-6) when a large portion of the layer is to be etched in a circuit design. Additionally, another plasma combination of chlorine with fluorine gas can produce deep vertical cuts through the film masking materials of a layer, as is often needed when producing microactuator sensor chips.

fluorine gas
氟气

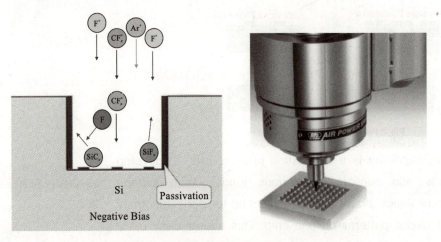

Fig. 21-5　Plasma Etching

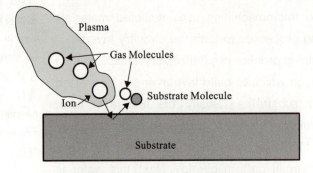

Fig. 21-6　Dry Plasma Etching

Tips

1. Silicon 硅

硅是一种化学元素，化学符号是 Si。其原子序数为 14，相对原子质量为 28.0855，有无定形硅和晶体硅两种同素异形体，位于元素周期表的第三周期，属于 ⅣA 族的类金属元素。硅也是极为常见的一种元素，然而它极少以单质的形式在自然界出现，而是以复杂的硅酸盐或二氧化硅的形式广泛存在于岩石、砂砾、尘土之中。硅主要用来制作高纯半导体、耐高温材料、光导纤维通信材料、有机硅化合物、合金等，被广泛应用于航空航天、电子电气、建筑、运输、能源、化工、纺织、食品、轻工、医疗、农业等行业。

New Words and Phrases

micromachining [ˌmaɪkrəʊməˈʃiːnɪŋ] *n*. 显微机械加工，微细加工
surface micromachining 表面微细加工
bulk micromachining 体微加工
ceramic [sɪˈræmɪk] *adj*. 陶瓷的
etching [ˈetʃɪŋ] *n*. 蚀刻
chemical etching 化学腐蚀法
wet etching 湿蚀刻法
plasma [ˈplæzmə] *n*. 等离子体
cantilever gate electrode 悬臂栅电极
drain bias voltage 漏极偏置电压
drain diffusion 泄漏扩散
polarization voltage 极化电压
silicon substrate 硅基板
reagent [riˈeɪdʒənt] *n*. 试剂
resultant [rɪˈzʌltənt] *n*. 生成物
boundary layer 边界层
fabrication [ˌfæbrɪˈkeɪʃn] *n*. 制作，制造，建造，装配
fabrication process 制造工艺
crystalline [ˈkrɪstlaɪn] *adj*. 水晶（般）的

microscopic [ˌmaɪkrəˈskɒpɪk] *adj*. 微观的
macroscopic [ˌmækrəˈskɒpik] *adj*. 宏观的
dissolve [dɪˈzɒlv] *v*. （使）溶解，解散，解除，消失
intact [ɪnˈtækt] *adj*. 原封不动的，完整的
vaporize [ˈveɪpəraɪz] *v*. （使）蒸发
hazardous [ˈhæzədəs] *adj*. 危险的，有害的，碰运气的
gyroscope [ˈdʒaɪərəskəʊp] *n*. 陀螺仪
substrate [ˈsʌbstreɪt] *n*. 基片
silicon oxide 二氧化硅
insulator [ˈɪnsjʊleɪtə] *n*. 绝缘体
ultraviolet（UV）light 紫外线
overlay [ˌəʊvəˈleɪ] *n*. 覆盖，涂层 *v*. 覆在……上面，覆盖，铺
rinse [rɪns] *v*. 冲洗
electrolyze [ɪˈlektrəʊlaɪz] *v*. 电解
ionize [ˈaɪənaɪz] *v*. （使）电离，离子化
bombard [bɒmˈbɑːd] *v*. 冲击，轰击
motion sensor 运动传感器
accelerometer [əkˌseləˈrɒmɪtə(r)] *n*. 加速度计，加速器
vehicle airbag accelerometer 车辆安全气囊加速计
navigation system gyroscope 导航系统陀螺仪
plastic substrate 塑料基板
fluorine [ˈflɔːriːn] *n*. 氟
fluorine gas 氟气

Notes

1. as opposed to 而不是

例句：Bulk micromachining starts with a solid piece and removes material until it reaches its final shape, as opposed to surface micromachining, which builds a piece layer by layer.

微细加工是指将一块固体去除材料直至其最终形状，不像表面微细加工那样逐层构建。

例句：This is a book about business practice as opposed to business theory.

这是一本关于商务实践而非理论的书。

2. break down into 分解为

例句：In addition, it possesses a very fine crystalline structure that can break down into layers thinner than human hair.

此外,它的晶体结构非常精细,可以分解成比人类头发更薄的层。

例句:After many years, rocks break down into the dirt.

经过许多年,岩石分解成泥土。

3. be incorporated into 使成为……的一部分

例句:This allows many more functions and electronics to be incorporated into each chip for use in motion sensors, vehicle airbag accelerometers, or for use in navigation system gyroscopes.

这使得每个芯片可以集成更多的功能和电子器件,以用于运动传感器、车辆安全气囊加速计,或用于导航系统陀螺仪。

例句:Undertakings of physical culture and sports shall be incorporated into the plan for the national economy and social development.

体育事业应当纳入国民经济和社会发展计划。

4. in combination with 结合

例句:Dry etching can be done alone, or in combination with chemical etching, using an ionized gas to bombard the areas to be etched.

干法蚀刻可以单独使用或结合化学蚀刻使用,使用电离气体冲击蚀刻区域。

例句:The data are employed in combination with cloud maps.

这些资料与云图结合一起使用。

Exercises

Ⅰ. Write True or False beside the following statements about the text.

1. _____ Micromachining doesn't use wafers of silicon, but will occasionally use plastic or ceramic material.

2. _____ The most common method for performing bulk micromachining is via selective masking and dry chemical solvents.

3. _____ The vast majority of bulk micromachining uses silicon.

4. _____ The most common method for bulk micromachining is called dry chemical etching.

5. _____ Manufacturers use dry plasma etching when a large portion of the layer is to be etched in a circuit design.

Ⅱ. Answer the following questions in English according to the text.

1. What is the most common method for performing bulk micromachining?

2. What are the differences between the dry etching process and the wet chemical etching process?

3. What is surface micromachining?

Ⅲ. Read the text again and fill in the blanks in the following sentences orally.

1. Micromachining is a method of making extremely _____ mechanical or electrical components.

2. Machine complex systems of gears and levers are also typical examples of the _____ of micromachining.

3. The work piece is exposed to a solvent, which will then _____ any unprotected areas and leave the rest _____.

4. The surface micromachining process uses either _____ _____ chip substrates as a foundation upon which to build layers, or can be started on cheaper glass or plastic _____.

5. Additionally, another plasma _____ of chlorine with fluorine gas can produce deep _____ cuts through the film masking materials of a layer, as is often needed when producing microactuator sensor chips.

Ⅳ. Translation.

The most common method for bulk micromachining is called wet chemical etching. First, the workpiece is covered in a material that will protect it from a selected solvent. The protective mask is then selectively removed to expose the areas of the piece that are coming off. The workpiece is exposed to a solvent, which will then dissolve any unprotected areas and leave the rest intact. Afterward, the remaining masking material is removed.

 Lesson 22   Laser Processing

In industrial applications, two basic laser types are used for materials processing—the CO_2 laser (gas state)(as shown in Fig. 22-1) and the Nd:YAG laser (solid-state) (as shown in Fig. 22-2). Both laser types generate light in the infrared range of the spectrum, which means that the laser beam itself is not visible. The operator must take adequate safety precautions[①].

Margin Note

in the infrared range of
在红外线范围内

1. Laser Cutting

Laser cutting plays a dominant role in[②] metal processing today. High power beam sources, fiber delivery, and highly automated peripheral material handling systems are the reasons for high cutting speeds, extreme accuracy, and high productivity in both 2D and 3D applications, as shown in Fig. 22-3.

fiber delivery
纤维输送

peripheral material
外围材料

Axial Gas Flow Laser

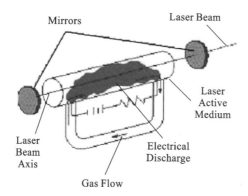

Fig. 22-1   Axial Gas Flow Laser

Nd:YAG Laser

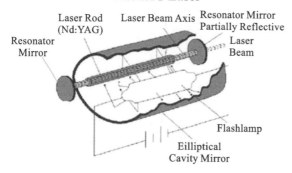

Fig. 22-2   Nd:YAG Laser

Fig. 22-3   Laser Cutting

When cutting with oxygen, the material is burned and vaporized after being heated to ignition temperature by a laser beam. The reaction between the oxygen and the metal produces additional energy in the form of heat, which supports the cutting process. When cutting thicker plate materials, it is important to reduce the flow of oxygen since otherwise, violent reactions will occur and reduce the cut quality. So in general, when oxygen is used as a cutting gas for mild steel, the rule of thumb[3] is: the thicker the material the lower the flow of gas. Cutting

ignition
temperature
燃点

with oxygen leaves a layer of oxides on the cut edge and can cause a reduction of alloying elements.

When cutting with nitrogen (or any other inert gas), the material is melted solely by the laser power and blown out of the cut kerf by the kinetic energy of the gas jet. As non-reactive gas does not react with the molten metal and does not generate④ additional heat, the laser power required is usually much higher than in oxygen of the same thickness. Also, the material does not overheat due to exothermic reactions. This means higher pressure is needed to expel the molten metal out of the cut kerf. Therefore, in nitrogen cutting, higher pressure is required as the thickness of the material increases.

Cutting with nitrogen is often used for aluminum and stainless steel to achieve good edge quality and maintain corrosion resistance. However, it is also applied to carbon steels whenever subsequent painting or powder coating is required.

2. Laser Welding

Laser welding (as shown in Fig. 22-4) is increasingly used in industrial products ranging from microelectronics to shipbuilding, and automotive manufacturing. It has proven the most outstanding in applications that take advantage of⑤ the many benefits of this technology:

(1) Low heat input.
(2) Small heat-affected zone (HAZ).
(3) Low distortion rate.
(4) High welding speed.

Fig. 22-4　Laser Welding

In welding applications, to focus on the high CO_2 laser powers, water-cooled mirrors are primarily used instead of lenses. The welding

alloying element
合金元素

kinetic energy
动能

non-reactive gas
非反应性气体

exothermic reaction
放热反应

stainless steel
不锈钢
corrosion resistance
耐蚀性
carbon steel
碳素钢

water-cooled mirror
水冷反射镜

is carried out in two ways. In conductive welding, the heat is transferred from the surface into the material via thermal conduction. This is typical for low-power Nd:YAG laser welding with relatively shallow welds. High-power laser welding is characterized by keyhole welding. The laser energy then melts and evaporates the metal. The pressure of the vapor displaces the molten metal so that a cavity is formed—the keyhole. The keyhole supports the transfer of the laser energy into the metal and guides the laser beam deep into the material. Keyhole welding thus allows very deep and narrow welds to be obtained and is therefore also called deep penetration welding.

Welding gas plays an important role in laser welding and fulfills several demands—the shielding of the weld pool and the heat-affected area, protection of the optics against fumes and spatter, root protection, and "plasma control" during CO_2 laser welding.

Welded plasma is ionized metal vapor and gas that can form the keyhole above. This cloud affects laser radiation and thus has the potential to interrupt the welding process.

thermal conduction
热传导

keyhole welding
穿透型焊接法
deep penetration welding
深熔焊

plasma control
等离子体控制

ionized metal vapor
电离金属蒸气

New Words and Phrases

infrared [ˌɪnfrə'red] n. 红外线 adj. 红外线的
visible ['vɪzəbl] adj. 明显的,看得见的
adequate ['ædɪkwət] adj. 充足的,适当的,胜任的
precaution [prɪ'kɔːʃ(ə)n] n. 预防,警惕,预防措施
dominant ['dɒmɪnənt] adj. 显性的,占优势的,支配的,统治的
role [rəʊl] n. 角色,任务
ignition [ɪg'nɪʃ(ə)n] n. 点火,点燃,着火
ignition temperature 燃点
violent ['vaɪəl(ə)nt] adj. 暴力的,猛烈的
reaction [rɪ'ækʃ(ə)n] n. 反应
oxygen ['ɒksɪdʒ(ə)n] n. 氧气
oxide ['ɒksaɪd] n. 氧化物
nitrogen ['naɪtrədʒ(ə)n] n. 氮气
inert gas 惰性气体
kinetic [kɪ'netɪk] adj. 运动的,活跃的

kinetic energy 动能
overheat [ˌəʊvəˈhiːt] vi. 过热，愤怒起来 vt. 使过热，使愤怒 n. 过热，激烈
exothermic [ˌeksə(ʊ)ˈθɜːmɪk] adj. 发热的，放热的
exothermic reaction 放热反应
expel [ɪkˈspel] vt. 驱逐，开除
molten metal 熔融金属
subsequent [ˈsʌbsɪkw(ə)nt] adj. 后来的，随后的
powder [ˈpaʊdə] n. 粉，粉末
shipbuilding [ˈʃɪpbɪldɪŋ] n. 造船，造船业
distortion rate 失真率
lens [ˈlenz] n. 透镜
thermal [ˈθɜːm(ə)l] adj. 热的，热量的
thermal conduction 热传导
evaporate [ɪˈvæpəreɪt] vt. 使……蒸发，使……脱水，使……消失
vapor [veɪpə] n. 蒸汽，烟雾
optic [ˈɒptɪk] adj. 光学的，视觉的，眼睛的
cavity [ˈkævɪtɪ] n. 腔，洞，凹处
plasma control 等离子体控制
ionized metal vapor 电离金属蒸气
in the infrared range of 在红外线范围内
fiber delivery 纤维输送
peripheral material 外围材料
alloying [əˈlɔɪɪŋ] n. 合金化处理，炼制合金
alloying element 合金元素
non-reactive gas 非反应性气体
steel [stiːl] n. 钢，钢铁
stainless steel 不锈钢
carbon steel 碳素钢
corrosion [kəˈrəʊʒ(ə)n] n. 腐蚀，侵蚀
corrosion resistance 耐蚀性
water-cooled mirror 水冷反射镜
keyhole welding 穿透型焊接法
penetration [ˌpenəˈtreɪʃ(ə)n] n. 穿透，渗透
deep penetration welding 深熔焊

 *Notes*

1. take precaution 采取预防措施,未雨绸缪

例句:The operator must take adequate safety precautions.
操作者必须采取足够的安全预防措施。

例句:It's important to prepare for your trip in advance and to take precautions while you are traveling.
在旅行时,提前做好准备是很重要的。

2. play a dominant role in 在……中占主要地位

例句:Laser cutting plays a dominant role in metal processing today.
如今激光切割在金属加工中扮演着主要角色。

例句:Expert decision plays a dominant role in the WTO dispute settlement mechanism.
专家决策在WTO争端解决机制中起着主导作用。

3. the rule of thumb 经验法则

例句:The rule of thumb is:the thicker the material the lower the flow of gas. Cutting with oxygen leaves a layer of oxides on the cut edge and can cause a reduction of alloying elements.
经验法则是:材料越厚氧气流量越低。用氧气切割会使切缝处产生降低合金元素含量的氧化层。

例句:The rule of thumb in the field is that each recorded minute of interaction takes an hour to analyze.
该领域依经验得出的规则是:每记录一分钟的互动都要用一个小时的时间来分析研究。

4. generate 生成,发生,产生

例句:As non-reactive gas does not react with the molten metal and does not generate additional heat.
因为非反应性气体不与熔化的金属发生反应,所以也就不产生多余的热量。

例句:The consortium stresses, though, that power generated by solar fields in North Africa would be used by North Africans as well as Europeans.
然而,该财团强调,在北非太阳能电厂生产的电力将提供给欧洲和北非。

5. take advantage of 利用

例句:It has proven the most outstanding in applications that take advantage of the many benefits of this technology.
事实证明,它是最出色的应用,利用了这项技术的许多优点。

例句：We should take advantage of these good conditions.
我们应充分利用这些良好的条件。

 Exercises

Ⅰ. Write True or False beside the following statements about the text.

1._____ When cutting with oxygen, the material is burned and vaporized before being heated up to ignition temperature by the laser beam.

2._____ When cutting thicker plate materials, it is important to increase the oxygen flow since otherwise, violent reactions will occur and reduce the cut quality.

3._____ In welding applications, to focus the high CO_2 laser powers, water-cooled mirrors are primarily used instead of lenses.

4._____ The welding is carried out in three ways.

5._____ The weld plasma is a cloud of ionized metal vapor and gases that can be formed in the keyhole.

Ⅱ. Answer the following questions in English according to the text.

1. What are the basic laser types used for material processing?
2. What is the rule of thumb when oxygen is used as a cutting gas for mild steel?
3. What is weld plasma?

Ⅲ. Read the text again and fill in the blanks in the following sentences orally.

1. Both laser types _____ light in the infrared range of the spectrum, which means that the laser beam itself is not _____.

2. When cutting with oxygen, the material is _____ and _____ after being heated to ignition temperature by a laser beam.

3. As non-reactive gas does not _____ _____ the molten metal and does not generate additional heat, the laser power required is usually much higher than in oxygen of the same thickness.

4. This means higher pressure is needed to _____ the molten metal out of the cut kerf.

5. Therefore, in nitrogen cutting, higher pressure is required as the thickness of the _____ _____.

Ⅳ. Translation.

When cutting with nitrogen (or any other inert gas), the material is melted solely by the laser power and blown out of the cut kerf by the kinetic energy of the gas jet. As non-reactive gas does not react with the molten metal and does not generate additional heat, the laser power required is usually much higher than in oxygen of the same thickness. Also, the material does not overheat due to exothermic reactions. This means higher pressure is needed to expel the molten metal out of the cut kerf. Therefore, in nitrogen cutting, higher pressure is required as the thickness of the material increases.

Lesson 23   EDM Machine

Every machinist knows that on standard machine tools, electrical energy is converted into motion by an electric motor. Nowadays, it has been discovered that electrical energy can be directly employed in metal removal. Electrical discharge machining (EDM) (as shown in Fig. 23-1), or spark machining, as it is so called, is based on the eroding effect of an electric spark on both the electrodes used to produce it. The second important discovery is that the rate of erosion by the spark is greatly increased if it is made to take place① in liquid rather than air. The main advantage is that electrical discharge machining processes with good surface finish and can machine metal of any hardness.

Margin Note

metal removal
金属切削
eroding effect
侵蚀效应

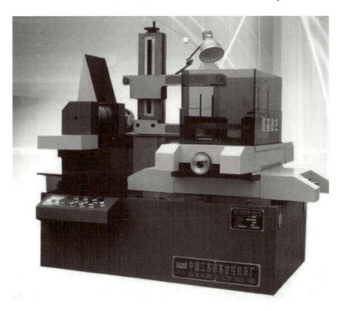

Fig. 23-1　Electrical Discharge Machining

EDM is a form of metal removal in which pulsating direct current is applied to a shaped tool (electrode) and a workpiece, both of which are capable of② conducting electricity. The two are held close to each other with a nonconducting fluid serving as an insulator between them. When a voltage high enough to break down the insulator is reached, a spark

nonconducting fluid
非导电流体

jumps the gap between the tool and the workpiece. As a result, this spark removes a small portion of the material. The simplified diagram in Fig. 23-2 illustrates the principle of EDM. The tool is mounted on the chuck attached to[③] the machine spindle whose vertical feed is controlled by the servomotor through a reduction gearbox. The workpiece is placed in a tank filled with a dielectric fluid; a depth of at least 50 mm over the work surface is maintained to eliminate the risk of fire. The tool and workpiece are connected to a DC relaxation circuit fed from a DC generator. Dielectric fluid is circulated under the pressure by a pump, usually through a hole (or holes) in the tool electrode. The servomotor maintains a spark gap of about 0.025 to 0.05 mm. It is maintained at a constant value by the servomechanism that actuates and controls the tool feed.

vertical feed
垂直给进
reduction gearbox
减速箱
dielectric fluid
绝缘液体
DC relaxation circuit
直流弛张电路
constant value
恒定值

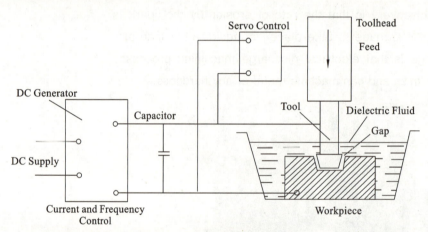

Fig. 23-2　The Principle of EDM

Electrical discharge machining is independent of[④] the mechanical properties of the workpiece. Any metal, regardless of[⑤] hardness, strength, or structure, can be machined. Almost unlimited geometries are possible as well[⑥], and cutting forces are zero. On the other hand, temperatures generated along the spark channel may reach several thousand degrees Fahrenheit. Therefore, dielectric fluid may have to be circulated at relatively large rates to prevent metallurgical changes.

metallurgical change
金相组织变化

New Words and Phrases

electrical [iˈlektrik(ə)l] *adj*. 电的, 与电有关的
electrical energy 电能
convert [kənˈvɜːt] *v*. 转换,(使)转变, 转化, 可转变为, 可变换成
electric motor 电动机
removal [riˈmuːvəl] *n*. 移走, 去掉, 消除, 清除
metal removal 金属切削
discharge [disˈtʃɑːdʒ] *v*. 放电
electrical discharge machining (EDM) 电火花加工
spark [spɑːk] *n*. 火花, 火星, 诱因
spark machining 电火花加工
electrode [iˈlektrəud] *n*. 电极
erosion [iˈrəuʒən] *n*. 腐蚀(作用), 磨损
finish [ˈfiniʃ] *v*. 完成, 做好 *n*. 修整, 精修, 精整, 精制, 修正
hardness [ˈhɑːdnis] *n*. 硬度, 刚度, 强度,(坚)硬性, 刚度指数
pulsate [pʌlˈseit] *v*. 脉动, 搏动, 跳动, 波动, 振动, 颤动, 抖动
shaped tool 成型刀具
conducting [ˈkɔndʌktiŋ] *adj*. 传导的 *n*. 指挥, 传导
conducting electricity 感应电流
fluid [ˈfluː(ː)id] *n*. 流体, 液体
nonconducting fluid 非导电流体
voltage [ˈvəultidʒ] *n*. 电压
gap [gæp] *n*. 间隙, 间隔, 间断, 空白, 空隙, 距离, 范围
servomotor [ˈsɜːvəu.məutə] *n*. 伺服电机
reduction gearbox 减速箱
tank [tæŋk] *n*.(液体、气体、储藏)容器
dielectric [ˌdaiiˈlektrik] *adj*. 非传导性的, 诱电性的
dielectric fluid 绝缘液体
maintain [menˈtein] *v*. 维护, 保养, 支持, 保持
risk of fire 火灾危险
relaxation [riːlækˈseiʃən] *n*. 松弛的, 弛张的
DC relaxation circuit 直流弛张电路
generator [ˈdʒenəreitə] *n*. 发电机, 产生者
DC generator 直流发电机
circulate [ˈsɜːkjuleit] *v*. 循环,(使)环行, 环流
pressure [ˈpreʃə(r)] *n*. 压力 *v*. 对……施加压力

pump [pʌmp] *n*. 泵，抽水机 *v*. 抽吸
constant value 恒定值
strength [streŋθ] *n*. 强度，浓度，力量
cutting force 切削力
temperature [ˈtempritʃə(r)] *n*. 温度
spark channel 电花通道
Fahrenheit [ˈfærənhait] *n*. 华氏温度（计）（的），华氏（温标）
metallurgical [metəˈlɜːdʒikəl] *adj*. 冶金的，冶金学的
metallurgical change 金相组织变化

 Notes

1. take place 发生，举行

例句：The second important discovery is that the rate of erosion by the spark is greatly increased if it is made to take place in liquid rather than air.

另一个重要发现是，如果电火花发生在液体中而不是空气中，其腐蚀速度将会大大提高。

例句：The next meeting will take place on Thursday.

下一次会议将在星期四举行。

2. be capable of 有才能的，有能力的，可容纳的，有技能的，有资格的

例句：EDM is a form of metal removal in which pulsating direct current is applied to a shaped tool (electrode) and a workpiece, both of which are capable of conducting electricity.

电火花加工是将脉冲直流电施加于可导电的成型工具（电极）和工件上，从而切除金属的一种方式。

例句：He is capable of judging art.

他具有艺术鉴赏能力。

3. attach to 附属，加入，参加

例句：The tool is mounted on the chuck attached to the machine spindle whose vertical feed is controlled by the servomotor through a reduction gearbox.

工具安装在附在机床主轴的卡盘上做垂直进给运动，通过减速箱由伺服电机控制。

例句：Prof. Smith was attached to the medical college as a guest professor for two years.

史密斯教授在医学院当了两年的客座教授。

4. be independent of 不依赖……，独立于……

例句：Electrical discharge machining is independent of mechanical properties of workpiece.

电火花加工与工件的力学性能无关。

例句:The reports from two separate sources are entirely independent of one another.
这两份独立来源的报告完全互不相干。

5. regardless of 不管,无论,不顾

例句:Any metal, regardless of hardness, strength or structure, can be machined.
任何金属无论其硬度、强度和结构如何均能被加工。

例句:We will persevere regardless of past failures.
尽管以前我们失败过,但仍要坚持下去。

6. as well 同时,倒不如,还是……的好,最好……还是

例句:Almost unlimited geometries are possible as well, and cutting forces are zero.
同时,几乎无限的几何形状是可能的,并且切削力为零。

例句:The weather was so bad that we might just as well have stayed at home.
天气糟糕透了,还不如待在家里。

Exercises

Ⅰ. Write True or False beside the following statements about the text.

1. _____ Electrical energy can be directly used in cutting metals.

2. _____ Spark machining is another name for EDM.

3. _____ In EDM, both shaped tools and workpieces are capable of conducting electricity.

4. _____ Electrical discharge machining is independent of the mechanical properties of the workpiece.

5. _____ Spark machining is faster in the air than in liquid.

Ⅱ. Answer the following questions in English according to the text.

1. What is electrical discharge machining based on?

2. What does the diagram in Fig. 23-2 illustrate?

3. Why may dielectric fluid have to be circulated at relatively large rates?

Ⅲ. Read the text again and fill in the blanks in the following sentences orally.

1. Every machinist knows that on standard machine tools, _____ _____ is converted into motion by an _____ _____.

2. The second important discovery is that the rate of erosion by the _____ is greatly increased if it is made to take place in _____ rather than _____.

3. When a _____ high enough to break down the _____ is reached, a spark jumps the gap between the tool and the workpiece.

4. The workpiece is placed in a tank filled with a _____ _____; a depth of at

least 50 mm over the work surface is maintained to eliminate the risk of _____.

5. Therefore, dielectric fluid may have to be circulated at relatively large rates to prevent _____ _____.

IV. Translation.

Electrical discharge machining is independent of the mechanical properties of the workpiece. Any metal, regardless of hardness, strength, or structure, can be machined. Almost unlimited geometries are possible as well, and cutting forces are zero. On the other hand, temperatures generated along the spark channel may reach several thousand degrees Fahrenheit. Therefore, dielectric fluid may have to be circulated at relatively large rates to prevent metallurgical changes.

Lesson 24 Additive Manufacturing

Additive manufacturing (AM), sometimes known as three-dimensional printing, is a process of making a three-dimensional solid object of virtually any shape from a digital model by adding layer-upon-layer upon of material.

The first successful attempts at additive manufacturing came from technology developed in the 1970s, though additive's earliest roots can be traced to① topography and photo sculpture, both first developed in the 1800s.

Common to AM technologies is the use of computers, three-dimensional modeling software, machine equipment, and layering material. Once a CAD sketch is produced, the AM equipment reads the data in the CAD file and places or adds successive layers of liquid, powder, sheet material, or others, in a layer-upon-layer manner to fabricate a three-dimensional object.

The term AM encompasses many technologies including subsets like three-dimensional printing, rapid prototyping (RP), direct digital manufacturing (DDM), layered manufacturing, and additive fabrication.

AM applications are limitless. Early use of AM in the form of rapid prototyping focused on the preproduction of visualization models. More recently, AM is being used to fabricate end-use products in aircraft, dental restorations, medical implants, automobiles, and even fashion products. While the adding of layer-upon-layer approach is simple, there are many applications of AM technologies with degrees of sophistication

Margin Note

three-dimensional
三维的

layer-upon-layer
层层

direct digital manufacturing
直接数字化制造

end-use product
终端产品
dental restoration
牙齿修复
medical implant
医学植入

to meet diverse needs including:

(1) A visualization tool in design.

(2) A means to create highly customized products② for consumers and professionals alike③.

(3) An industrial tooling.

(4) To produce small lots of④ production parts.

industrial tooling
工业工具

(5) One day.... production of human organs.

There are several different categories and processes of additive manufacturing available, each appropriate for different materials and requirements. These are detailed in the diagrams below.

Powder bed processes (as shown in Fig. 24-1) consolidate thin layers of powder using a laser or electron beam to fuse scans of the sliced computer-aided design (CAD) data to create the geometry. A recoater mechanism is used to lay down the powder on top of each scanned area, allowing you to build up the part layer by layer.

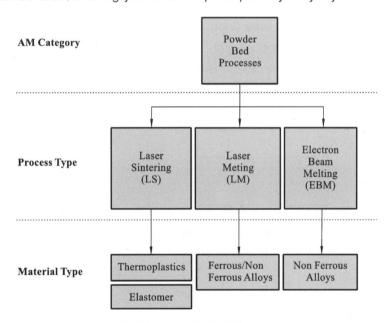

Fig. 24-1 Powder Bed Processes

Material deposition processes (as shown in Fig. 24-2) work by heating the material through an extrusion nozzle which follows a predefined deposition path, layering on top of a platform, and depositing material on top of previous layers to create the three-dimensional object. The material hardens immediately upon extrusion from the nozzle.

extrusion nozzle
挤压喷嘴

Three-dimensional printing (as shown in Fig. 24-3) works by laying

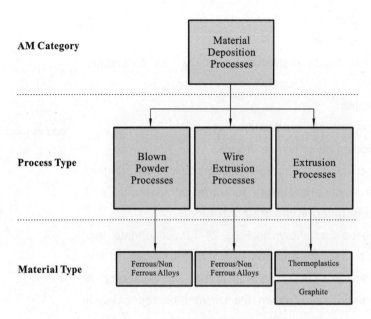

Fig. 24-2　Material Deposition Processes

down thin layers of heated material onto a platform. Either the head or platform will be continuously moving to deposit more material on top of each other to form the three-dimensional object. Binders and powder can also be used to form three-dimensional objects. Three-dimensional printing is faster, more affordable, and easier to use than other additive technologies.

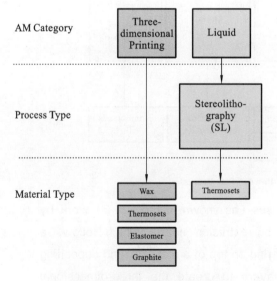

Fig. 24-3　Three-dimensional Printing

　　The liquid vat process solidifies thin layers together, using an ultraviolet (UV) curable thermoset polymer liquid with a solid state

liquid vat process
液体还原法

crystal laser to create the required geometry layer by layer, using computer-aided design (CAD) data. A recoater mechanism is used to cover the previous layer with the material enabling the next layer to be scanned. Three-dimensional on platform is shown in Fig. 24-4.

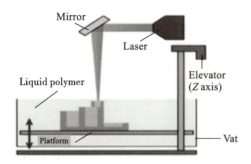

Fig. 24-4　Three-dimensional Printing on Platform

 New Words and Phrases

virtually [ˈvɜːtjuəli] *adv*. 事实上
topography [təˈpɒgrəfi] *n*. 地形学
photo sculpture 照相雕刻法
sketch [sketʃ] *n*. 素描, 草图
successive [səkˈsesiv] *adj*. 继承的, 连续的
subset [ˈsʌbset] *n*. 子集
prototyping [ˈprəʊtəʊtaɪpɪŋ] *n*. 原型设计
rapid prototyping (RP) 快速原型
direct digital manufacturing (DDM) 直接数字化制造
visualization [ˌvɪʒuəlaɪˈzeɪʃn] *n*. 形象化, 可视化
restoration [ˌrestəˈreɪʃn] *n*. 恢复, 复位
implant [imˈplɑːnt] *n*. 植入物
consolidate [kənˈsɒlɪdeɪt] *vt*. 巩固, 合并, 统一
electron [iˈlektrɒn] *n*. 电子
fuse [fjuːz] *v*. 熔化, 融合
geometry [dʒiˈɒmətri] *n*. 几何图形, 几何结构
deposition [ˌdepəˈzɪʃn] *n*. 沉淀, 罢免, 废黜
extrusion [ikˈstruːʒn] *n*. 挤压
predefined [ˌpriːdɪˈfaɪnd] *adj*. 预定义的
binder [ˈbaɪndə] *n*. 黏结剂

vat [væt] n. 大桶
solidify [səˈlidifai] v. 使凝固
ultraviolet [ˌʌltrəˈvaiələt] adj. 紫外的
thermoset [ˈθɜːməuset] adj. 热固性的
thermoset polymer 热固性聚合物

Notes

1. be traced to 追溯到

例句：The first successful attempts at additive manufacturing came from technology developed in the 1970s, though additive's earliest roots can be traced to topography and photo sculpture, both first developed in the 1800s.

增材制造的第一次成功尝试来自20世纪70年代发展的技术，尽管增材制造技术最早可以追溯到地形测量技术和照相雕刻法，这两种技术最初都发展于19世纪。

例句：The definition of meter can be traced to the source via the step height in the surface measurement technology.

在表面测量技术中，测量仪的定义可以通过台阶高度来溯源。

2. customized product 定制产品

例句：A means to create highly customized products for consumers and professionals alike.

一种为消费者和专业人士创造高度定制产品的手段。

例句：However, customized products will bring more pressure on the company and raise more new requirements.

然而，定制产品将会给企业带来更大的压力并提出新的要求。

3. alike 一样地

例句：A means to create highly customized products for consumers and professionals alike.

一种为消费者和专业人士创造高度定制产品的手段。

例句：Brisk daily walks are still the best exercise for young and old alike.

不管是年轻人还是老人，每天快步行走仍是最好的锻炼方式。

4. small lots of 小批量

例句：To produce small lots of production parts.

生产小批量的部件。

例句：Inspectors are responsible for the review of small lots of non-conforming products in the production process.

检验员负责生产过程中小批量不合格产品的评审。

Exercises

Ⅰ. Write True or False beside the following statements about the text.

1. _____ The first successful attempts at additive manufacturing came from technology developed in the 1980s.

2. _____ Early use of AM in the form of rapid prototyping focused on the preproduction of visualization models.

3. _____ There are not many applications of AM technologies with degrees of sophistication to meet diverse needs.

4. _____ Neither the head nor the platform will be continuously moving to deposit more material on top of each other to form the 3D object.

5. _____ 3D printing is faster, more affordable, and easier to use than other additive technologies.

Ⅱ. Answer the following questions in English according to the text.

1. What is additive manufacturing?
2. What is AM being used to fabricate?
3. What did early use of AM in the form of rapid prototyping focus on?

Ⅲ. Read the text again and fill in the blanks in the following sentences orally.

1. Additive manufacturing (AM), sometimes is known as _____ _____.

2. The first successful _____ at additive manufacturing came from technology developed in the 1970s.

3. The term AM _____ many technologies including subsets like three-dimensional printing.

4. More recently, AM is being used to fabricate end-use products in aircraft, dental restorations, _____ _____, _____, and even _____ _____.

5. Powder bed processes _____ thin layers of powder using a laser or electron beam to fuse scans of the sliced computer-aided design (CAD) data to create the _____.

Ⅳ. Translation.

Common to AM technologies is the use of computers, three-dimensional modeling software, machine equipment, and layering material. Once a CAD sketch is produced, the AM equipment reads the data in the CAD file and places or adds successive layers of liquid, powder, sheet material, or others, in a layer-upon-layer manner to fabricate a three-dimensional object.

Lesson 25　Manual Programming Training of CNC Machining Center

　　There are two kinds of machining center programming: manual programming and automatic programming. This lesson introduces manual programming. The drawing of machined parts is shown in Fig. 25-1. It is required to write the part processing program correctly and complete the processing to obtain qualified parts.

Margin Note

manual programming
手工编程
automatic programming
自动编程

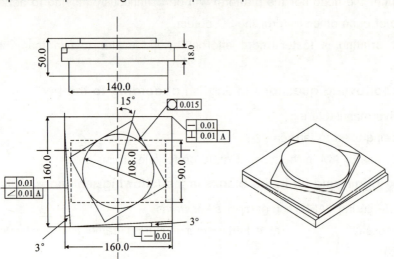

Fig. 25-1　Drawing of Machined Parts

1. The Tasks

The technical description is as follows.

（1）Material: 2A12 modulated aluminum, or equivalent.

（2）All dimensions are free tolerances (±0.5 mm).

（3）Unspecified sharp angle 0.5×45°.

（4）Processing by a CNC machine tool: the processed surface is flat, and the finish is shown in the Fig. 25-1.

　①Processing: φ108 mm circle, two 3° beveled edges, 108 mm×108 mm 15° beveled, 160 mm×160 mm square.

　②160 mm×160 mm square. The milling depth of four sides is greater than 6 mm, and all upper planes do not need processing.

　③Ensure accuracy according to the geometric tolerance shown in the Fig. 25-1.

modulated aluminum
调制铝
free tolerance
自由公差

milling depth
铣切深度
geometric tolerance
形位公差

2. Introduction to Basic Instructions

1) G90, G91

(1) G90. The absolute value instruction performs linear cutting to the specified point based on the difference between the specified point and the program zero point.

(2) G91. The incremental value instruction performs linear cutting to the specified point based on the difference between the specified point and the starting point.

Fig. 25-2 shows an example.

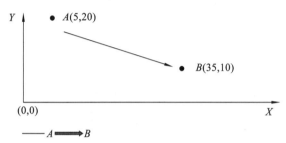

Fig. 25-2   Example 1

The instructions are shown below:

G90 G00 X35 Y10

G91 G00 X30 Y-10

2) G00, G01

(1) G00. Format: G00 X_ Y_ Z_; (X, Y, Z: specified point coordinates).

With the command of straight-line rapid positioning, each axis moves quickly to the specified point without cutting at the shortest distance, and the speed is generally set to① 48 000 mm/min.

(2) G01. Format: G01 X_ Y_ Z_ F_; (X, Y, Z: specified point coordinates; F: feed speed, mm/min).

Linear cutting instructions are the commands to move from point A to point B with the shortest distance between two points to complete the path.

The instructions are shown below:

G90 G00 X5 Y20

G01 X35 Y10 F1000

3) G02, G03

(1) Format: G02/G03 X_ Y_ R_ (I_ J_) F_.

Clockwise/anticlockwise arc interpolation instruction enables the tool to move along an arc on the specified plane.

X: Moving distance of the X axis or its parallel axis (arc end coordinate).

Y: Moving distance of the Y axis or its parallel axis (arc end coordinate).

I: Distance from the starting point of the X-axis to the center point of the arc (vector of the center of the circle relative to the starting point).

vector 矢量

J: Distance from the starting point of the Y-axis to the center point of the arc (vector of the center of the circle relative to the starting point).

R: Radius of arc (with a symbol, >180° is "−"; <180° is "+").

Fig. 25-3 shows an example.

Fig. 25-3 Example 2

The instructions are shown below:

From "A" to "B"
G02 X-10 Y0 I-10 F2000
From "B" to "A"
G03 X10 Y0 I10 F2000
From "A" to "A"
G03 X10 Y0 I-10 F2000

(2) Points to note:

①When making arc cutting, the position of the spindle should first run to the starting point of the arc.

②The instruction format is G02 (or G03) X_ Y_ I_ (or J_)F_.

③X and Y are arc end coordinates.

④I and J are the coordinates of the center point of the arc minus the coordinates of the starting point of the arc.

4) G40, G41, G42

Format: G40. Tool radius correction is canceled.

Format: G41 D_. Tool radius left compensation.

Format: G42 D_. Tool radius right compensation.

Because of the existence of tool diameter, its influence on the path must be considered when programming. To process the correct contour potential, it is necessary to[2] make the tool offset a tool radius towards the appropriate orientation. This offset action is called a correction.

tool offset
刀具偏置

The tool radius can be calculated manually during programming, and the correction can also be realized through the recognized codes G41 and G42 of the machine itself.

D_: Indicates the tool diameter correction number, such as D1.

Fig. 25-4 shows an example.

Use the G41 command to compensate the left side of the clockwise movement of the tool path, which enlarges the size while tooling the workpiece or diminutive the size while tooling the corner

Use the G41 command to compensate the right side of the anticlockwise movement of the tool path, which diminutive the size while tooling the workpiece or enlarges the sice while tooling the corner

Fig. 25-4   Example 3

5) G43, G44, G49

Format: G43 H_ Z_, tool length positive complement; G44 H_ Z_, tool length negative complement (hardly used).

Function: Use the Z-axis position to correct the error of tool length (as shown in Fig. 25-5).

Fig. 25-5   Example 4

H_: indicates the tool complement number.

Z_: the end coordinate of the Z axis usually refers to[3] the safe altitude.

G49: tool length correction canceled.

Format: G49.

6) G54-G59

G54-G59: workpiece coordinate system.

G90 G10 L2 P_ (P1-P6) X_ Y_ Z_ G54: workpiece coordinate system 1 Select P1.

workpiece coordinate system
工件坐标系

G55: workpiece coordinate system 2 Select P2.
G56: workpiece coordinate system 3 Select P3.
G57: workpiece coordinate system 4 Select P4.
G58: workpiece coordinate system 5 Select P5.
G59: workpiece coordinate system 6 Select P6.

7) G68, G69

Format: G68 X_ Y_ R_.

Function: coordinate rotation, all rotation movement will be around the rotation center.

X_ Y_: absolute coordinates of the rotation center.

R_: rotation angle (P in wattage system).

Example: G68 X0 Y0 R90 (coordinate rotation 90 degrees).

G69: coordinate rotation canceled.

Such as A workpiece programming can be processed into A, A1, or A2 (as show in Fig. 25-6).

wattage system 瓦数系统
coordinate rotation 坐标轴旋转

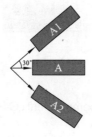

Fig. 25-6　A Workpiece Programming be Processed into A, A1, or A2

A1: coordinate rotation with 30 degree.
A2: coordinate rotation with −30 degree.

8) M00, M01, M02, M30, M99

(1) M00: program pause.

Function: when the CNC executes this instruction, the spindle will stop rotating and the feed will stop.

(2) M01: selective program pause.

Function: when CNC executes this command and this switch is on, it has the same function as M00.

(3) M02: program ends.

Function: after the processing of the main program, if you want to re-execute the program, you need to press the "reset" key to return the program to the beginning.

(4) M30: program ends.

Function: indicates that the program ends here and returns to the beginning of④ the program.

(5)M99:program ends.

Function:circular instructions, subroutine end return to the main program. The program loop.

9)M03,M04,M05,M06,M08,M09

(1)M03:spindle turning forward.

Function:spindle clockwise rotation(M03 S_).

(2)M04:spindle reversal.

Function:the main shaft rotates counterclockwise(M04 S_).

(3)M05:spindle stop.

Function:spindle stops rotating.

(4)M06:tool change instruction.

Function:execute tool exchange instruction(M06 T_).

(5)M08:cutting fluid on.

Function:open cutting fluid.

(6)M09:cutting fluid off.

Function:close cutting fluid.

New Words and Phrases

aluminum [əˈluːmɪnəm] n. 铝

modulated [ˈmɔdjuleitid] adj. 已调的,被调的

modulated aluminum 调制铝

compensation [ˌkɒmpenˈseɪʃ(ə)n] n. 弥补,抵消,赔偿金,补偿金

anticlockwise [ˌæntiˈklɒkwaɪz] adj. 逆时针的

beveled edge 斜边

clockwise [ˈklɒkwaɪz] adj. 顺时针的

clockwise/anticlockwise arc interpolation 顺时针/逆时针圆弧插补

wattage [ˈwɒtɪdʒ] n. 瓦数

wattage system 瓦数系统

description [dɪˈskrɪpʃn] n. 说明,描述,形容

geometric tolerance 形位公差

incremental [ˌɪŋkrəˈment(ə)l] adj. 增加的,递增的,逐渐的

incremental value instruction 增量值指令

contour [ˈkɒntʊə(r)] n. 轮廓,外形,周线

manual programming 手工编程
angle [ˈæŋg(ə)l] n. 倾斜,斜角,角度
milling depth 铣切深度
interpolation [ɪnˌtɜːpəˈleɪʃ(ə)n] n. 插入,篡改,填写,插值
absolute [ˈæbsəluːt] adj. 绝对的,无疑的
absolute value instruction 绝对值指令
linear [ˈlɪnɪə(r)] adj. 线性的,直线的
linear cutting instruction 直线切削指令
arc [ɑːk] n. 弧形,弧线
offset [ˈɒfset] v. 使偏离直线方向 n. 偏离量,偏离距离
tool offset 刀具偏置
coordinate [kəʊˈɔːdɪneɪt] n. 坐标 v. 协调,配合
coordinate rotation 坐标轴旋转
plane [pleɪn] n. 平面
feed speed 进给速度
radius [ˈreɪdɪəs] n. 半径
command [kəˈmɑːnd] n. (计算机的)指令,命令,指示
rotation [rəʊˈteɪʃn] n. 旋转,转动
straight-line rapid positioning 直线快速定位
technical [ˈteknɪk(ə)l] adj 工艺的,技术的
automatic programming 自动编程
free tolerance 自由公差
workpiece coordinate system 工件坐标系

Notes

1. set to 调到

例句:With the command of straight-line rapid positioning, each axis moves quickly to the specified point without cutting at the shortest distance, and the speed is generally set to 48 000 mm/min.

通过直线快速定位指令,各轴以最短的距离在无切削状态下快速移动至指定点,速度一般设为 48 000 mm/min。

例句:All macros can be set to run when a hotkey is pressed.
所有宏都可以被设定为按下某一快捷键时开始运行。

2. it is necessary to… 必须

例句:To process the correct contour potential, it is necessary to make the tool offset a tool radius towards the appropriate orientation.

要加工出正确的轮廓势必要使刀具向着合适的方位偏移一个刀具半径。

例句：It is necessary to examine how the proposals can be carried out.

有必要调查一下怎样才能实施这些方案。

3. refer to 指的是，适用于，涉及

例句：The end coordinate of the Z axis usually refers to the safe altitude.

Z 轴终点坐标通常是指安全高度。

例句：You know who I'm referring to.

你知道我指的是谁。

4. the beginning of ……的开始

例句：Function：indicates that the program ends here and returns to the beginning of the program.

功能：表示程序到此结束，并返回到程序最开始的位置。

例句：Can I go back to what you said at the beginning of the meeting?

我想回到你在会议开始时所提的话题，行吗？

Exercises

Ⅰ. Write True or False beside the following statements about the text.

1. _____ It is required to write the part processing program correctly.

2. _____ G90, the incremental value instruction, performs linear cutting to the specified point based on the difference between the specified point and the starting point.

3. _____ When making arc cutting, the position of the spindle should first run to the ending point of the arc.

4. _____ To process the correct contour potential, it is necessary to make the tool offset a tool radius towards the appropriate orientation.

5. _____ M06 means to execute the tool exchange instruction.

Ⅱ. Answer the following questions in English according to the text.

1. What are the types of CNC machining centers?

2. What is the specific content of the technical description?

3. What is called correction? How can you realize the correction?

4. Describe the function of the following instructions：M00, M30, M99?

Ⅲ. Read the text again and fill in the blanks in the following sentences orally.

1. All dimensions are _____ _____.

2. Processing by a CNC machine tool：the processed surface is _____.

3. With the command of _____ _____ positioning, each axis moves quickly to the specified point without cutting at the shortest distance.

4. _____ _____ instructions are the commands to move from point *A* to point *B* with the shortest distance between two points to complete the path.

5. _____/_____ arc interpolation instruction enables the tool to move along an arc on the specified _____.

Ⅳ. Translation.

Because of the existence of tool diameter, its influence on the path must be considered when programming. To process the correct contour potential, it is necessary to make the tool offset a tool radius towards the appropriate orientation. This offset action is called a correction.

Reading Material 17   3D Printing Technology

3D printers use standard inkjet printing technology to create parts layer by layer by depositing a liquid binder on thin layers of powder. Instead of feeding paper under the print heads like a 2D printer, a 3D printer moves the print heads to powder bed, where it prints the cross-sectional data sent from the printing software. The printing system requires powder to be distributed accurately and evenly across the building platform. 3D printers accomplish this by using a feed piston and platform, which rises incrementally for each layer. A roller mechanism spreads powder feed from the feed piston to the building platform, as shown in Fig. R17-1, intentionally spreading approximately 30 percent additional powder on each layer to ensure a full layer of densely packed powder on the building platform. The excess powder falls down an overflow chute into a container and is reused in the next build.

Once the layer of powder is spread, the inkjet print heads print the cross-sectional area for the first, or bottom slice of the part onto the smooth layer of powder, binding the powder together. A piston then lowers the building platform by 0.101 6 mm, and a new layer of powder is spread on top. The print heads apply the data for the next cross-section onto the new layer, as shown in Fig. R17-2, which binds itself to the previous layer. The printing system repeats this process for all of the layers of the part. The 3D printing process creates an exact physical model of the geometry represented by 3D data. Process time depends on the height of the part or parts being built. Typically, our 3D printers have a vertical speed of 25-50 mm per hour.

Margin Note

inkjet
喷墨
liquid binder
液态黏合剂
cross-sectional data
横截面数据
feed piston
馈料活塞
roller mechanism
辊机构
overflow chute
溢流槽

cross-sectional area
横截面面积

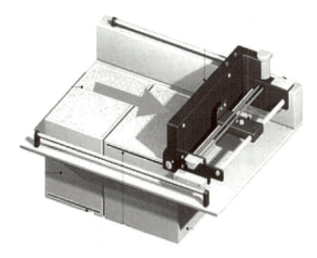

Fig. R17-1 Spread a Layer of Power

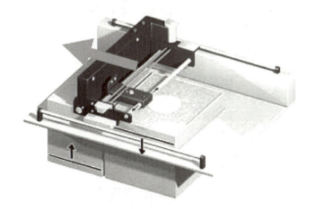

Fig. R17-2 Print Cross-section

When the 3D printing process completes, loose powder surrounds and supports the part in the building chamber. Users can remove the part from the building chamber after the materials solidify, and return unprinted, loose powder to the feed platform for reuse. Users then use forced air to blow the excess powder off the printed part, a short process that takes less than 10 minutes. The 3D printing technology does not require the use of solid or attached supports during the printing process, and all unused materials are reusable.

Our 3D printing technology is the fastest additive technology commercially available on the market. Other companies often refer to their equipment as 3D printers, however, these systems rely on processes using a vector approach or single-jet technology to deposit all building materials. We use inkjet print heads with a resolution of 600 dpi (dots per inch), focuses on a drop-on-demand approach, and

vector approach
矢量方法
single-jet technology
单射流技术
resolution
分辨率
drop-on-demand approach
随需应变的方法

manufacture the only true 3D inkjet printers available. The technology allows the printing of multiple parts simultaneously, while only adding a negligible amount of time to the print time for one part. Many people mistakenly believe that "raster is faster but vector is corrector", but this is not always the case. In printing, especially 3D printing, the accuracy of the model depends on the ability to jet when and where required. This is a function of jet size and motion control. Our technology's precise inkjet implementation results in high definition and high quality parts.

raster
光栅

Also contributing to the overall speed of the 3D printing process is the method used to distribute the material. We use a spreading method for depositing more than 90 percent of the material, which is extremely efficient and fast. Our 3D printers dispense only a small percentage of the building material, called binders, through the print heads. Other additive prototyping technologies deposit 100 percent of the building material through a nozzle, resulting in very slow printing speeds and lengthy prototype turnaround times.

prototyping technology
原型技术
nozzle
喷嘴
turnaround time
周转时间

 New Words and Phrases

incrementally [ˌɪnkrɪˈmentəli] adv. 递增地，逐渐地，增加地
densely [ˈdensli] adv. 稠密地，密集地
deposit [dɪˈpɒzɪt] v. 放置，使沉积，使沉淀
negligible [ˈneɡlɪdʒəbl] adj. 可以忽略不计的，微不足道的，不重要的
raster [ˈræstər] n. 光栅
dispense [dɪˈspens] v. 分配，分发，施与
additive [ˈædətɪv] adj. 附加的，加成的，加法的
nozzle [ˈnɒzl] n. 喷嘴
inkjet [ˈɪŋkˌdʒet] n. 喷墨
liquid binder 液态黏合剂
cross-sectional data 横截面数据
feed piston 馈料活塞
roller mechanism 辊机构
overflow chute 溢流槽
cross-sectional area 横截面面积
single-jet technology 单射流技术

resolution [ˌrezəˈluːʃ(ə)n] n. 分辨率
drop-on-demand approach 随需应变的方法
prototyping technology 原型技术
turnaround time 周转时间

Reading Material 18　Nanotechnology

Nanolithography (NL) is a growing field of techniques within nanotechnology dealing with the engineering (patterning e.g. etching, depositing, writing, printing, etc) of nanometer scale structures on various materials.

The modern term reflects on a design of structures built in the range of 10^{-9} to 10^{-6} meters, i.e. nanometer scale. Essentially, the field is a derivative of lithography, only covering very small structures. All NL methods can be categorized into four groups: photolithography (as shown in Fig. R18-1), scanning lithography, soft lithography, and other miscellaneous techniques.

As of 2021 photolithography is the most widely used technology in the mass production of microelectronics and semiconductor devices. It's characterized by both high production throughput and small-sized features of the patterns.

Optical lithography (or photolithography) (as shown in Fig. R18-2) is one of the most important and popular technologies in the nanolithography field. Optical lithography contains several important derivative techniques, all of that use very short light wavelengths to change the solubility of certain molecules, causing them to wash away in solution, leaving behind a desired structure. Several optical lithography techniques require the use of liquid immersion and a host of resolution enhancement technologies like phase shift mask (PSM) and optical proximity correction (OPC). Some of the included techniques in this set include multiphoton lithography, X-ray lithography, light coupling mask (LCM) nanolithography, and extreme ultraviolet lithography (EUVL). This last technique is considered to be the most important next-generation lithography (NGL) technique due to its ability to produce structures accurately down below 30 nanometers at high throughput rates which makes it a viable option for commercial purposes.

Margin Note
nanolithography
纳米光刻
scanning lithography
扫描光刻
soft lithography
软光刻
semiconductor device
半导体器件
production throughput
生产吞吐量

optical lithography
光学光刻

liquid immersion
液浸
resolution enhancement technology
分辨率增强技术
phase shift mask
相移掩模

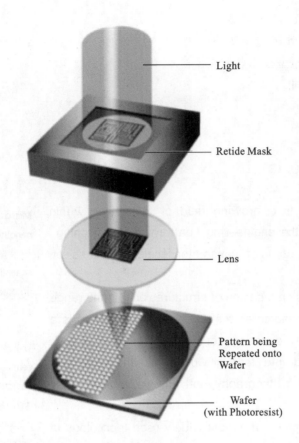

Fig. R18-1　Photolithography

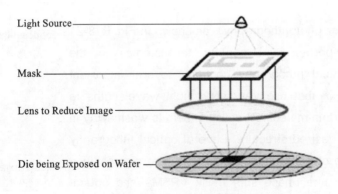

Fig. R18-2　Optical Lithography

Quantum optical lithography (QOL), is a diffraction-unlimited method able to write at 1 nm resolution by optical means, using a red laser diode ($\lambda = 650$ nm). Complex patterns like geometrical figures and letters were obtained at 3 nm resolution on a resist substrate. The method was applied to nanopattern graphene at 20 nm resolution.

optical proximity correction 光学邻近校正
multiphoton lithography 多光子光刻
X-ray lithography X射线光刻
light coupling mask nanolithography 光耦合掩模纳米光刻
extreme ultraviolet lithography 极紫外光刻
throughput rate 通量率

quantum optical lithography 量子光学光刻
resist substrate 抗蚀剂衬底
nanopattern graphene 纳米图形石墨烯

Electron-beam lithography (EBL) or electron-beam direct-write (EBDW) lithography scans a focused beam of electrons on a surface covered with an electron-sensitive film or resist (e. g. PMMA or HSQ) to draw custom shapes. By changing the solubility of the resist and subsequent selective removal of material by immersion in a solvent, sub-10 nm resolutions have been achieved. This form of direct-write, maskless lithography has high resolution and low throughput, limiting single-column e-beams to photomask fabrication, low-volume production of semiconductor devices, and research & development. Multiple-electron beam approaches have as a goal an increase of throughput for semiconductor mass production. EBL can be utilized for selective protein nanopatterning on a solid substrate, aimed at ultrasensitive sensing.

Scanning probe lithography (SPL) is another set of techniques for patterning at the nanometer scale down to individual atoms using scanning probes, either by etching away unwanted material or by directly writing new material onto a substrate. Some of the important techniques in this category include dip-pen nanolithography, thermochemical nanolithography, thermal scanning probe lithography, and local oxidation nanolithography. Dip-pen nanolithography is the most widely used of these techniques.

electron-beam lithography 电子束光刻
electron-beam direct-write lithography 电子束直写光刻
electron-sensitive film 电子敏感膜
custom shape 自定义形状
single-column e-beam 单列电子束
photomask fabrication 光掩模制造
multiple-electron beam 多电子束
ultrasensitive sensing 超灵敏传感
scanning probe lithography 扫描探针光刻
dip-pen nanolithography 蘸笔纳米光刻
thermochemical nanolithography 热化学纳米光刻
thermal scanning probe lithography 热扫描探针光刻
local oxidation nanolithography 局部氧化纳米光刻

New Words and Phrases

etch [etʃ] v. 蚀刻,凿出
miscellaneous [ˌmɪsəˈleɪniəs] adj. 各种各样的,混杂的
viable [ˈvaɪəbl] adj. 可行的,可实施的,能独立发展的
lithography [lɪˈθɒɡrəfi] n. 刻蚀术
photolithography [ˌfəʊtəlɪˈθɒɡrəfi] n. 光刻
scanning lithography 扫描光刻
soft lithography 软光刻
semiconductor device 半导体器件
production throughput 生产吞吐量
optical lithography 光学光刻

solubility [ˌsɒljuˈbɪləti] n. 溶解度
liquid immersion 液浸
resolution enhancement technology 分辨率增强技术
phase shift mask 相移掩模
optical proximity correction 光学邻近校正
multiphoton lithography 多光子光刻
X-ray lithography X射线光刻
light coupling mask nanolithography 光耦合掩模纳米光刻
extreme ultraviolet lithography 极紫外光刻
throughput rate 通量率
quantum optical lithography 量子光学光刻
diffraction [dɪˈfrækʃn] n. 衍射
resist substrate 抗蚀剂衬底
nanopattern graphene 纳米图形石墨烯
electron-beam lithography 电子束光刻
electron-beam direct-write lithography 电子束直写光刻
electron-sensitive film 电子敏感膜
resist [rɪˈzɪst] n. 抗蚀剂
custom shape 自定义形状
single-column e-beam 单列电子束
photomask fabrication 光掩模制造
multiple-electron beam 多电子束
ultrasensitive sensing 超灵敏传感
scanning probe lithography 扫描探针光刻
dip-pen nanolithography 蘸笔纳米光刻
thermochemical nanolithography 热化学纳米光刻
thermal scanning probe lithography 热扫描探针光刻
local oxidation nanolithography 局部氧化纳米光刻

Reading Material 19 High-speed Machining

The recent advances in machining technology are the practice of high-speed machining (HSM) (as shown in Fig. R19-1). HSM is a metal-cutting process that emphasizes rapid speed and feed rates to enhance productivity and surface quality. HSM is widely used in aerospace production, mold manufacturing, automotive industries, micromachining, precision components, optic industries, and household appliances.

Margin Notes
high-speed machining
高速加工
metal cutting process
金属切削工艺
feed rate
进给速率

Fig. R19-1 High-speed Machining (HSM)

High-speed machining is important for manufacturing enterprises to maximize the use of CNC equipment. As companies grow, they need to process parts as quickly as possible while maintaining quality. Science-based tool paths allow users to increase production, extend tool life and reduce machine wear while continuing to use existing CNC machines and standard tools.

tool path
工具路径
machine wear
机床磨损

So how does it work? The secret lies in maintaining constant tool pressure during the cutting process, as shown in Fig. R19-2. It is accomplished by the tool path, cutter, and feed movement working together, which can be viewed as all instruments playing together to complete a favorite song. Traditional high-speed machining tool paths are just one constant feed rate through the cutting process, which creates a shock load on the tool, workpiece, and machine. These shocks over time lead to wear on the equipment and tooling. By varying the path of the cutting tool, the shock load can be reduced and the production time can be shortened at a lower cost.

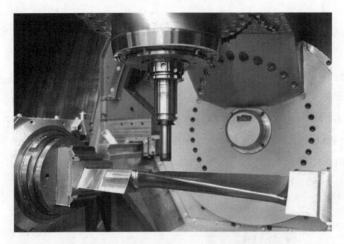

Fig. R19-2 Maintaining Constant Tool Pressure during the Cutting Process

SOLIDWORKS CAM allows you to streamline the manufacturing process by capturing standard workflows, reading MBD (model based definition) tolerances, and extending the life of tools and equipment, as shown in Fig. R19-3.

model based definition
基于模型的定义

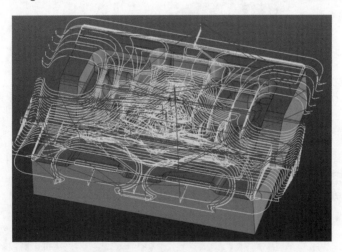

Fig. R19-3 SOLIDWORKS CAM

DATRON high-speed CNC machines produce precision, complex parts that are burr-free and surgically clean in a fraction of the time, as shown in Fig. R19-4. These compact CNC milling systems use less floor space than their large machining area, require less power than traditional machines, and can be customized to meet users' exact needs.

burr-free
无毛刺

milling system
铣削系统

DATRON also offers a complete line of universally compatible high-speed milling tools. If the machining speed is 15 000 rpm or above, it is

Fig. R19-4　High-speed CNC Machine Achieve Burr-free Cutting

more economical and effective to use high-speed cutting tools.

DATRON CNC machines and high-speed spindles will increase your feed rates and ensure faster cycle times. Since the material is fed faster, it can effectively evacuate the chips from the tool and the cutting channel to reduce friction and heat. Then you get a better part finish and longer cutting tool life.

Enhancing efficiency and part quality may allow you to take on tasks you couldn't do before, but the fastest CNC workflow comes from streamlining the entire production process from start to finish.

Faster milling: reduce your cycle times by as much as 400% with high-speed spindles and patented tooling technology.

Faster completion: no oil means the CNC machines complete parts with lightning speed and surgical cleaning.

Faster service: highly skilled service and support team are ready to quickly resolve any issues that may encounter.

Faster installation: installation is simpler, so the new DATRON CNC machine is up and running in hours.

Faster integration: DATRON machines are seamlessly integrated with the user's workflow, so the user makes parts almost immediately.

Faster setup: DATRON's touchscreen interface makes your work more efficient without CNC training.

touchscreen
触摸屏

New Words and Phrases

high-speed machining 高速加工
aerospace [ˈeərəuspeɪs] n. 航空航天工业/技术 adj. 航空和航天（工业）的
streamline [ˈstriːmlaɪn] v. 精简（组织、流程等）使效率更高,使……成为流线型
surface [ˈsɜːfɪs] n. 表面,水面,地面,桌面,台面
mold [məuld] n. 模具,铸模
fraction [frækʃn] n. 小部分,微量,分数,小数
chip [tʃɪp] n. 碎块,碎屑,芯片
surgically [ˈsɜːdʒɪkli] adv. 如外科手术般地
evacuate [ɪˈvækjueɪt] v. 疏散,撤离
friction [ˈfrɪkʃ(ə)n] n. 摩擦,摩擦力,不和,分歧
precision component 精密零件
feed rate 进给速率
tool path 工具路径
machine wear 机床磨损
model based definition 基于模型的定义
burr-free 无毛刺
milling system 铣削系统

Reading Material 20　Flexible Manufacturing System

The flexible manufacturing system（FMS）is a manufacturing system in which there is a certain amount of flexibility that allows the system to react in case of changes, whether predicted or unpredicted, as shown in Fig. R20-1 and Fig. R20-2.

Margin Note

flexible manufacturing system 柔性制造系统

Fig. R20-1　FMS in Factory

Fig. R20-2　FMS in Training Base

This flexibility is generally divided into two categories, both of which contain numerous subcategories.

The first category is called routing flexibility which covers the system's ability to be changed to produce new product types, and the ability to change the order of operations executed on a part.

The second category is called machine flexibility which consists of the ability to use multiple machines to perform the same operation on a part, as well as the system's ability to absorb large-scale changes, including volume, capacity, or capability.

Most FMSs consist of three main systems:

(1) Connect the work machine with an automated CNC machine.

(2) Optimize the part flow with the material handling system.

(3) Control material movements and machine flow by the central control computer.

The main advantage of an FMS is its high flexibility in managing manufacturing resources like time and effort to manufacture a new product. The best application of an FMS is found in the production of small sets of products such as batch production.

Flexibility in manufacturing means the ability to deal with slightly or greatly mixed parts, allow variations in parts assembly and process sequence, and change the production volume and the design of certain products being manufactured.

An industrial flexible manufacturing system (FMS) consists of robots, computer-controlled machines, computer numerically controlled (CNC) machines, instrumentation devices, computers, sensors, and other independent systems such as inspection machines. The use of robots in the production segment of manufacturing industries promises a variety of benefits ranging from high utilization to a high volume of productivity. Each robotic cell or node will be located along a material handlings system such as a conveyor or automatic guided vehicle. The production of each part or workpiece will require a different combination of manufacturing nodes. The movement of parts from one node to another is done through the material handling system. At the end of part processing, the finished parts will be routed to an automatic inspection node, and subsequently unloaded from the flexible manufacturing system.

The FMS data traffic consists of large files and short messages, mainly from nodes, devices, and instruments. The message size ranges from a few bytes to several hundreds of bytes. Executive software and other data, for example, are large files, while messages for machining data, instrument-to-instrument communications, status monitoring, and data reporting are transmitted in small sizes.

There are also some variations in response time. Large program files from the main computer usually take about 60 seconds to be downloaded into each instrument or node at the beginning of FMS operation. Messages for instrument data need to be sent at a periodic time with a deterministic time delay. Other types of messages used for emergency reporting are quite short in size and must be transmitted and received with an almost instantaneous response.

The demands for reliable FMS protocols that support all the FMS data characteristics are now urgent. Token bus has a deterministic message delay, but it does not support the prioritized access scheme which is needed in FMS communications. Token ring provides prioritized access and has a low message delay, but its data transmission is unreliable.

Since machine failure and malfunction caused by heat, dust, and electromagnetic interference are common, a prioritized mechanism and immediate transmission of emergency messages are needed so that a suitable recovery procedure can be applied. A modification is proposed to standardize the token bus to implement a prioritized access scheme to allow the transmission of short and periodic messages with a lower delay compared to the one for long messages.

automatic guided vehicle
自动导引车

data traffic
数据流

deterministic time delay
确定性时间延迟

instantaneous response
瞬间响应

token bus
令牌总线

token ring
令牌环

electromagnetic interference
电磁干扰

New Words and Phrases

subcategory [ˈsʌbˌkætəɡəri] n. 子范畴，亚类
node [nəʊd] n. 节点，结点，茎节
status [ˈsteɪtəs] n. 状况，情形
malfunction [ˌmælˈfʌŋkʃn] n. 故障，失灵
modification [ˌmɒdɪfɪˈkeɪʃn] n. 修改，更改
flexible manufacturing system 柔性制造系统
automatic guided vehicle 自动导引车
data traffic 数据流
deterministic time delay 确定性时间延迟
instantaneous response 瞬间响应
token bus 令牌总线
token ring 令牌环
electromagnetic interference 电磁干扰

Unit Ⅵ　Development Trend of Intelligent Manufacturing

   Introduction to the Unit 单元导言

> 随着移动互联、超级计算、大数据、云计算、物联网等新一代信息技术的飞速发展，新一轮工业革命聚焦在新一代人工智能技术的战略性突破上，呈现出深度学习、跨界协同、人机融合、群体智能等数字化、网络化和智能化特征，推动着制造业的产业升级。智能生产制造过程更加优质、柔性化、高效、绿色环保，极大提升了制造业的创新力和竞争力。本单元将重点讨论有关智能制造技术发展方面的内容。其主要内容包括敏捷制造、绿色制造、精益生产、计算机集成制造、协作机器人，以及机器学习、数字图像处理、最新的智能制造技术和新一代人工智能技术引领下的智能制造；目的是使学生在学习英语的同时，扩大专业知识面，成为具有综合素质、掌握最新知识和技能、具备初步国际从业能力的高端技术技能型人才。

   ## Lesson 26　Agile Manufacturing

Agile manufacturing is a professional term applied to[①] an organization that creates processes, tools, and employee training to respond quickly to customer needs and market changes while controlling costs and quality. It is mainly related to[②] lean manufacturing.

Agile manufacturing (as shown in Fig. 26-1) is a manufacturing method that focuses on[③] meeting customer needs while maintaining high-quality standards and controlling the overall costs of specific products. This approach is geared towards companies working in a highly competitive environment where minor variations in performance and product delivery can make a huge difference[④] in the company's long-term survival and reputation among consumers.

The concept is closely related to lean manufacturing, which aims to reduce waste as much as possible. In lean manufacturing, the company cuts all costs that are not directly related to consumers' products. In

Margin Note

agile manufacturing
敏捷制造

lean manufacturing
精益生产

addition, agile manufacturing means customer demands need to be fulfilled quickly and effectively. Combining the two concepts, they are sometimes referred to as using "agile and lean manufacturing". Companies utilizing agile manufacturing methods tend to have formidable network connections with suppliers and partner companies as well as numerous cooperative teams working within companies to deliver products efficiently. They can also quickly update equipment, negotiate new agreements with suppliers and other partners in response to⑤ market changes, and take appropriate measures for customer requirements. Therefore, the company will increase the production of products with higher demand, and redesign products to solve problems in the market.

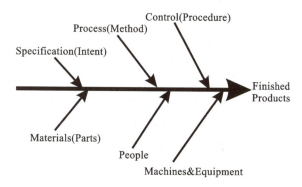

Fig. 26-1　Agile Manufacturing

The improvement of agility does not rely on high technology and massive investment. However, proper technology and advanced management can help enterprise agility to reach new heights.

1. Information Assurance Technical Framework (IATF)

To deal with market challenges and support the operation of dynamic alliances, we need a cross-enterprise, cross-industry, and cross-regional information technology framework to support the collaborative operation of different enterprises. In this way, we realize the remote cooperation within and between multifunctional groups in a heterogeneous distribution environment.

dynamic alliance
动态联盟
heterogeneous
distribution
environment
异构分布环境

2. Design Model and Workflow Control System

A design model and a workflow control system (as shown in Fig. 26-2) are required to support integrated product process design. They cover serval areas, including the definition of the data model and process model of integrated products, the management of product data,

workflow
control system
工作流控制系统

dynamic resources, and development process during product development, and necessary centralized management of security measures and distributed systems.

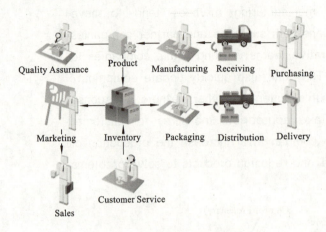

Fig. 26-2　Workflow Control System

security measure
安全措施
distributed system
分布式系统

3. Supply Chain Management System

The supply chain management system supports resource sharing and information integration among enterprises when the organization and operation mode of enterprises shift to agile enterprises and dynamic alliances.

supply chain management system
供应链管理系统

4. Agility of Various Equipment, Process, and Workshop Scheduling

We must improve the agility of all links of enterprises, thereby improving their agility greatly.

5. Agility Evaluation System

Agility evaluation is summarized into four indicators: time, cost, robustness, and adaptive range. Agile means fast. It means being good at handling challenges. Agile empowers businesses to seize opportunities at the right time and to lead the way through technological innovation. Thus, the agility of an enterprise is determined by its ability to grasp opportunities and achieve innovation.

New Words and Phrases

agile manufacturing 敏捷制造
respond [rɪˈspɒnd] v. 响应,(口头或书面)回答,做出反应,回应
lean [liːn] adj. 精简的,效率高的 v. 倚靠,靠在
lean manufacturing 精益生产
variation [ˌveərɪˈeɪʃn] n. 变化,变动
focus [ˈfəʊkəs] v. 集中,关注,聚焦,调焦
competitive [kəmˈpetətɪv] adj. 竞争的
minor [ˈmaɪnə(r)] adj. 较小的,次要的
delivery [dɪˈlɪvəri] n. 递送,投递,分娩,生产,演讲风格
reputation [ˌrepjuˈteɪʃn] n. 名誉,名声
demand [dɪˈmɑːnd] n. 需求,需求量 v. 强烈要求,需要,需求
environment [ɪnˈvaɪrənmənt] n. 周围状况,条件,工作平台,自然环境
negotiate [nɪˈɡəʊʃieɪt] v. 商定,达成协议,谈判,洽谈
partner [ˈpɑːtnə(r)] n. (合伙企业的)合伙人,(生意、组织或国家的)伙伴
design [dɪˈzaɪn] n. 设计,布局,安排
massive [ˈmæsɪv] adj. 大量的,大规模的
investment [ɪnˈvestmənt] n. 投资
cooperation [kəʊˌɒpəˈreɪʃn] n. 合作,协作
resource [rɪˈsɔːs] n. 资源,自然资源
adaptive [əˈdæptɪv] adj. 适应的,有适应能力的
seize [siːz] v. 抓住,捉住,把握(机会等)
grasp [ɡrɑːsp] v. 抓牢,握紧,理解,领会,抓住(机会)
survival [səˈvaɪv(ə)l] n. 存活,幸存
formidable [fəˈmɪdəb(ə)l] adj. (形状、数量)巨大的,可怕的,令人敬畏的
appropriate [əˈprəʊpriət] adj. 合适的,相称的
alliance [əˈlaɪəns] n. 联盟,结盟,(人们之间的)亲密关系,(尤指)联姻
dynamic alliance 动态联盟
heterogeneous [ˌhetərəˈdʒiːniəs] adj. 异质的,不均一的,参差的
heterogeneous distribution environment 异构分布环境
workflow control system 工作流控制系统
robustness [rəʊˈbʌstnəs] n. 稳健性,健壮性
supply chain management system 供应链管理系统

Notes

1. apply to 适用于,应用于

例句:Agile manufacturing is a professional term applied to an organization that creates processes, tools, and employee training.

敏捷制造是一个专业术语,适用于一个创建工艺、工具和员工培训的组织。

例句:The new technology was applied to farming.

这项新技术已应用于农业。

2. be related to 与……有关

例句:It is mainly related to lean manufacturing.

这主要与精益生产有关。

例句:In the future, pay increases will be related to productivity.

以后,工资的增加将和业绩挂钩。

3. focus on 专注于,集中于

例句:Agile manufacturing is a manufacturing method that focuses on meeting customer needs.

敏捷制造是一种专注于满足客户需求的制造方法。

例句:We shall focus on the needs of the customer.

我们将重点关注顾客的需求。

4. make a difference 有影响,有关系

例句:Minor variations in performance and product delivery can make a huge difference in the company's long-term survival and reputation among consumers.

性能和产品交付的微小变化可以对公司的长期生存和在客户中的声誉产生巨大的影响。

例句:These are young people who are making a difference in their communities, as well as on their teams.

这些年轻人正在改变他们的社区,以及他们的团队。

5. in response to 响应,回答,对……有反应

例句:They can also quickly update equipment, negotiate new agreements with suppliers and other partners in response to market changes, and take appropriate measures for customer requirements.

他们可以快速更新设备,与供应商和其他合作伙伴谈判新的协议,以应对市场的不断变化,并采取相应措施满足客户需求。

例句:The product was developed in response to customer demands.

这种产品是为了满足顾客的需求而开发的。

Exercises

Ⅰ. Write True or False beside the following statements about the text.

1. _____ Agile manufacturing is a term applied to an organization that creates processes, tools, and employee training.

2. _____ Agile manufacturing focuses on meeting customer needs while controlling the overall costs of all products.

3. _____ The company always increases the production of products that are in urgent demand and designs new products to replace old ones.

4. _____ The improvement of agility relies on high technology and massive investment.

5. _____ The evaluation of agility involves time, cost, location, and adaptive range.

Ⅱ. Answer the following questions in English according to the text.

1. What does agile manufacturing refer to?
2. What support can a good IT framework provide?
3. What are the main elements of the agility evaluation system?

Ⅲ. Read the text again and fill in the blanks in the following sentences orally.

1. Agile manufacturing is mainly related to _____ manufacturing.

2. Agile manufacturing focuses on meeting _____ needs while maintaining _____ standards and controlling the overall costs of _____ products.

3. Companies utilizing agile manufacturing methods tend to have _____ _____ _____ with suppliers and partner companies.

4. We realize the _____ cooperation within and between _____ groups in a heterogeneous distribution environment.

5. A design model and a workflow control system are required to support _____ product process design.

Ⅳ. Translation.

The concept is closely related to lean manufacturing, which aims to reduce waste as much as possible. In lean manufacturing, the company cuts all costs that are not directly related to consumers' products. In addition, agile manufacturing means customer demands need to be fulfilled quickly and effectively. Combining the two concepts, they are sometimes referred to as using "agile and lean manufacturing".

Lesson 27 Green Manufacturing

Green manufacturing is a professional term involving two aspects: first, the manufacturing of green products, particularly those used in renewable energy systems and various clean technology equipment; second, the greening of manufacturing which means reducing pollution, waste, and use of natural resources, recycling and reusing what was considered waste, and reducing emissions.

As shown in Fig. 27-1, the circular economy is an ecological economy. It requires the use of ecological laws to guide the economic activities of human society, which is characterized by low mining, high utilization, and low emission. All materials and energy should be used rationally and consistently in this ongoing economic cycle to reduce the impact of economic activities on the natural environment to as little as possible. There are three major principles of circular economy, that is, reduction, reuse, and recycling. Each principle is essential to the successful implementation of the circular economy.

Margin Note

green manufacturing
绿色制造
renewable energy system
可再生能源系统
ecological law
生态规律
circular economy
循环经济
scrap return
回炉料
linear economy
线性经济

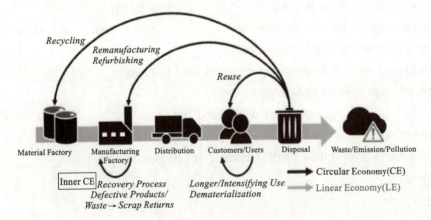

Fig. 27-1 Circular Economy

We have made great progress when it comes to[①] the greening of manufacturing. A growing number of people working in business realize the advantages of reduction in resource use, waste, and pollution, along with recycling and reusing what was formerly viewed as waste. To be specific, it not only increases the bottom line but also motivates employees, boosts their morale, and improves interpersonal

bottom line
最终盈利（或亏损）

relationships.

Corporate and business leaders are at the forefront of[2] redesigning, restructuring, re-engineering, and retooling equipment operations and manufacturing processes. They strive to achieve more sustainable development of the environment and society. They found that the corresponding results are measurable and others can emulate and hope to follow them, which even can lead to new business lines.

New ways of thinking about manufacturing, whether broadly or narrowly, will have a significant impact on[3] manufacturers globally. Such efforts are closely related to a trend that tends to assume or accept more corporate social responsibilities (CSR). Such a driving force has been the development of system analysis, which has gradually evolved into a growing field of industrial ecology.

Significantly, applying this systematic method to observe and analyze manufacturing technology and process will form a concept of product design and manufacturing which is called "from cradle to cradle". In brief, this method requires product design and manufacturing to focus on minimizing or even eliminating resource utilization, waste, and pollution, and emphasizes how to handle the product manufacturing process. The process is from the initial correct design of the product to the production, distribution, disposal, and reuse.

To promote their continuous progress, the government, international non-profit organizations, and business and industry regulators have urged manufacturers and enterprises to clean up their production and take greater responsibility for resource use, waste, and pollution.

Reduce the use of fossil fuels. Carbon dioxide and other greenhouse gas emissions have already been at the cutting edge of this movement. International organizations such as the United Nations and the World Bank are being united by a range of[4] public organizations, non-profit organizations, and private sector organizations, which have jointly promoted green production, green business methods, and green practices, and remarkable progress has been made. Prominent environmental protection organizations and enterprises have even bridged their differences and made their commitment to[5] protecting natural resources, natural habitats, and biodiversity.

re-engineering
流程再造

industrial ecology
工业生态学

carbon dioxide
二氧化碳

private
sector organization
私营部门组织

New Words and Phrases

professional [prəˈfeʃənl] adj. 专业的，职业的
term [tɜːm] n. 术语，专有名词
ecological [ˌiːkəˈlɒdʒɪkl] adj. 生态的，生态学的
ecological law 生态规律
mine [ˈmaɪn] v. 采（煤等矿物），挖掘 pron. 我的 n. 矿，矿井，宝库，源泉
utilization [ˌjuːtəlaɪˈzeɪʃn] n. 利用，使用
rationally [ˈræʃnəli] adv. 理性地
impact [ˈɪmpækt] n. 巨大影响，强大作用，撞击，冲击力
motivate [ˈməʊtɪveɪt] v. 激励，激发，成为……的动机
boost [buːst] v. 使增长，推动
morale [məˈrɑː] n. 士气，精神面貌
interpersonal [ˌɪntəˈpɜːsənl] adj. 人际关系的，人际的
strive [straɪv] v. 努力，力争
broadly [ˈbrɔːdli] adv. 广泛地，普遍地，大体上
minimize [ˈmɪnɪmaɪz] v. 使减小到最低限度，使最小化
greenhouse [ˈɡriːnhaʊs] n. 温室，暖房 adj. 温室效应的
remarkable [rɪˈmɑːkəbl] adj. 引人注目的，非凡的
renewable [rɪˈnjuːəbl] adj. 可再生的
emission [iˈmɪʃn] n. 排放物，散发物
reuse [ˌriːˈjuːz] v. 再次使用，重复使用
disposal [dɪˈspəʊzl] n. 处理，清除
waste [weɪst] n. 废物，废弃物 v. 浪费，滥用，未能充分利用
pollution [pəˈluːʃ(ə)n] n. 污染，污染物
forefront [ˈfɔːfrʌnt] n. （运动、活动的）前沿，（思考、关注的）重心
retool [ˌriːˈtuːl] v. 重组，重新装配
sustainable [səˈsteɪnəbl] adj. 可持续的
assume [əˈsjuːm] v. 承担，就职，假定，假设
dispose [dɪˈspəʊz] v. 处理，安排，丢弃
fossil fuel 化石燃料
greenhouse gas 温室气体（二氧化碳、甲烷等）
commitment [kəˈmɪtmənt] n. 承诺，保证，奉献，投入
habitat [ˈhæbɪtæt] n. 栖息地
biodiversity [ˌbaɪəʊdaɪˈvɜːsɪti] n. 生物多样性
green manufacture 绿色制造
renewable energy system 可再生能源系统

circular [ˈsɜːkjələ(r)] *adj*. 循环的,圆形的,环形的
circular economy 循环经济
linear economy 线性经济
bottom line 最终盈利(或亏损)
re-engineering 流程再造
industrial ecology 工业生态学
private sector organization 私营部门组织

Notes

1. when it comes to... 当提到,就……而论

例句:We have made great progress when it comes to the greening of manufacturing.
谈及制造的"绿化",我们已经取得巨大的进步。

例句:When it comes to pollution, the chemical industry is a major offender.
谈到环境污染问题,化工产业是一个"祸害"。

2. at the forefront of... 处于最前列

例句:Corporate and business leaders are at the forefront of redesigning, restructuring, re-engineering, and retooling equipment operations and manufacturing processes.
企业和商业领袖们处于重新设计、重组、流程再造,以及重建设备操作和制造工艺的最前沿。

例句:Women have always been at the forefront of the Green movement.
妇女总是在环境保护运动的最前列。

3. have an impact on 对……有影响

例句:New ways of thinking about manufacturing, whether broadly or narrowly, will have a significant impact on manufacturers globally.
关于制造的新思维方式,无论是广义还是狭义,都会在全球范围内对制造商产生重大影响。

例句:The presence or absence of clouds can have an important impact on temperature.
云的有无对气温会产生重要影响。

4. a range of 一系列

例句:International organizations such as the United Nations and the World Bank are being united by a range of public organizations, non-profit organizations, and private sector organizations.
一系列公众组织、非营利性组织和私营部门组织正在联合联合国和世界银行等国际组织。

例句：The hotel offers a wide range of facilities.

这家酒店提供各种各样的设施。

5. make a commitment to 承诺，献身

例句：Prominent environmental protection organizations and enterprises have even bridged their differences and made their commitment to protecting natural resources, natural habitats, and biodiversity.

著名的环保组织和企业甚至弥合分歧，致力于自然资源、自然栖息地和生物多样性的保护。

例句：The hospital makes a commitment to provide the best possible medical care.

这家医院承诺要提供最好的医疗服务。

Exercises

Ⅰ. Write True or False beside the following statements about the text.

1. _____ Greening manufacturing refers to reducing pollution and waste by minimizing natural resource use and waste.

2. _____ Green manufacturing is beneficial for businesses in terms of employee motivation, morale, and human relations.

3. _____ New ways of thinking about manufacturing are influencing manufacturers all over the world.

4. _____ Products should be designed and produced to maximize resource use, waste, and pollution in cradle-to-cradle manufacturing.

5. _____ Nowadays, it is only the government's job to drive the manufacturers to reduce pollution and waste.

Ⅱ. Answer the following questions in English according to the text.

1. What does green manufacturing refer to?

2. What is the circular economy?

3. What are the advantages of the greening of manufacturing?

4. What are the requirements for cradle-to-cradle?

Ⅲ. Read the text again and fill in the blanks in the following sentences orally.

1. The greening of manufacturing refers to reducing pollution, waste, and use of _____ resources, _____ and reusing what was considered waste, and reducing _____.

2. There are three major principles of circular economy: _____, reuse, and recycling.

3. Corporate and business leaders strive to achieve more _____ _____ of the environment and society.

4. To promote their continuous progress, the government, _____ _____ organizations, and bussiness and industry regulators have urged manufacturers and enterprises to clean up their production and take greater responsibility for resource use, waste, and pollution.

5. Prominent environmental protection organizations and enterprises have even bridged their _____ and made their _____ to protecting natural resources, natural habitats, and biodiversity.

Ⅳ. Translation.

The circular economy is an ecological economy. It requires the use of ecological laws to guide the economic activities of human society, which is characterized by low mining, high utilization, and low emission. All materials and energy should be used rationally and consistently in this ongoing economic cycle to reduce the impact of economic activities on the natural environment to as little as possible.

Lesson 28　Lean Production

Lean production, or lean manufacturing, refers to a business concept whose goal is to minimize the time and resources used in the manufacturing processes and other related activities of an enterprise as well as the response time of suppliers and customers, with an emphasis on① eliminating all forms of wastage. It is the fusion of various management philosophies designed to make operations as efficient as possible. The philosophies invoked by lean production include just in time (JIT) manufacturing, Kaizen, total quality management (TQM), total productive maintenance (TPM), cellular manufacturing (CM), etc.

Margin Note

lean production
精益生产

cellular manufacturing
单元式制造

Lean manufacturing follows three principles:

(1) Loss and waste are harmful.

(2) The manufacturing processes must be closely tied to the market's requirements.

(3) A company should be regarded as② a unified continuum including its customers and suppliers, which is called the "value stream".

value stream
价值流

Lean manufacturing is not only a method but also a way of life that all members of an organization must understand and practice.

The basic elements of lean production are:

(1) Timely and efficient production based on the principle of continuous product flow (also known as single-piece workflow).

(2) Continuous improvement of processes throughout the value chain, especially the quality and cost.

(3) To build multi-functional and multi-skilled teams at all levels to achieve their goals. Lean manufacturing is essentially a 21st century update on the 20th century theory of "mass production".

Among the elements mentioned above, the most prominent should be "continuous product flow", which requires rearranging the production workshop so that products are gradually processed from one workstation to another with minimal waiting and processing time between workstations. It means that the whole process route is dedicated to[③] processing a batch of similar products, or the products are subject to[④] a similar process. The equipment and workbenches are streamlined to ensure production continuity and efficiency. Attention should also be paid to[⑤] machine maintenance, uptime, and equipment utilization. Such a manufacturing setup is also known as cellular manufacturing.

In line with lean production, the following forms of waste should be eliminated: standby, temporary inventory, inventory transportation, overproduction, overprocessing, unnecessary actions, and defective devices. These wastes can be reduced in a production workshop that conforms to a continuous product flow. Another strategy is to focus the production on customer orders, which means only products that are immediately needed by customers (or the next workstation) should be produced. Thus, the site that needs to process inventory should be the site that extracts the inventory from the previous site.

Kaizen is a major influence on lean production. That's why lean manufacturing promotes teamwork among multi-skilled, multi-functional individuals at all levels, resulting in continuous process improvement of zero inventories, zero downtime, paperless, zero defects, and zero latency throughout the organization.

Cellular manufacturing refers to a manufacturing system in which equipment and workstations are arranged in an efficient sequence so that inventory and consumable materials run continuously and smoothly from beginning to end in a single process flow, while minimizing shipping or standby time, or without any delay. Cellular manufacturing is an

---

continuous product flow
连续产品流
single-piece workflow
单件工作流
mass production
批量生产

standby
待机

temporary inventory
临时库存

zero inventory
零库存
zero downtime
零停机
zero latency
零时延

important part of lean production.

In order to create a single process flow (or single product flow) production line, it is necessary to put all equipment for product processing in the same area. It differs from the setting in traditional mass production where only similar equipment is placed in the same area. According to the setting, when a process requires specific equipment to process parts, the parts need to be transported to the area where the certain equipment is located. They then line up there for batch processing, which can sometimes lead to delays in the delivery and processing of materials.

Tips

1. JIT: Just In Time 准时制生产方式

准时制生产方式,又称为无库存生产方式(stockless production)、零库存(zero inventories)、一个流(one-piece flow)或者超级市场生产方式(supermarket production)。JIT 的本质是保持材料流和信息流在生产过程中的同步,从而实现在必要时以必要的数量生产或供应必要的产品。

2. Kaizen 经营方法改善

Kaizen 是一个日语词汇,意指小的、连续的、渐进的改进。Kaizen 思想所带来的哪怕是细微效果,其结果往往是颠覆性的、革命性的。它要求每一位人员以相对较少的费用来不断地改进工作。长期而言,这种持续进步可以获得巨大的回报。Kaizen 也是低风险的方式。

3. TQM: Total Quality Management 全面质量管理

全面质量管理是指一个组织以质量为中心、以全员参与为基础,目的在于通过顾客满意及本组织所有成员和社会受益来实现永续发展的管理途径。在全面质量管理中,质量这个概念和全部管理目标的实现有关。

New Words and Phrases

invoke [ɪnˈvəʊk] v. 引起,提及,调用,激活
continuum [kənˈtɪnjuəm] (pl. continua) n. 连续体
methodology [ˌmeθəˈdɒlədʒi] n. 方法,原则,制备工艺

cellular manufacturing 单元式制造
value stream 价值流
continuous [kən'tɪnjuəs] adj. 连续不断的,持续的
continuous product flow 连续产品流
single-piece workflow 单件工作流
temporary ['temprəri] adj. 临时的,短暂的
temporary inventory 临时库存
zero inventory 零库存
zero downtime 零停机
zero latency 零时延

Notes

1. emphasis on 着重于,对……的强调

例句:Lean production aims to minimize the time and resources used in the manufacturing processes and other related activities of an enterprise as well as the response times of suppliers and customers, with an emphasis on eliminating all forms of wastage.

精益生产旨在将企业制造过程和其他相关活动中使用的时间和资源以及供货商和用户的响应时间减至最少,重点是消除所有形式的损耗。

例句:We provide all types of information, with an emphasis on legal advice.

我们提供各种信息服务,尤其是法律咨询。

2. be regarded as 被认为是……

例句:A company should be regarded as a unified continuum including its customers and suppliers, which is called the "value stream".

公司应被视为一个统一连续的整体,包括其用户和供货商,这一概念被称为"价值流"。

例句:He is regarded as a tenacious and persistent interviewer.

他被认为是一位执着而坚毅的采访者。

3. dedicate to 致力于

例句:It means that the whole process route is dedicated to processing a batch of similar products.

这意味着将整条工艺路线专用于加工一批相似的产品。

例句:She vowed to herself that she would dedicate her life to scientific studies.

她默默地发誓要献身于科学研究。

4. be subject to 使服从,受……管制

例句:The products are subject to a similar process.

产品接受一组相似的加工工艺。

例句：Prices may be subject to alteration.

价格可能会受变更影响。

5. pay attention to 注意，重视

例句：Attention should also be paid to machine maintenance, uptime, and equipment utilization.

还应该注意机器的维护、正常运行时间和设备利用率。

例句：I didn't pay attention to what she was saying.

我没有注意她在说什么。

 Exercises

Ⅰ. Write True or False beside the following statements about the text.

1. _____ Lean production aims at maximizing the time and resources used in the manufacturing processes.

2. _____ Lean manufacturing follows three principles.

3. _____ To achieve goals, lean production builds multi-functional and multi-skilled teams at some levels.

4. _____ Continuous product flow rearranges the production workshop so that products are gradually processed from one workstation to another with minimal waiting and processing time.

5. _____ Cellular manufacturing has never been an important part of lean production.

Ⅱ. Answer the following questions in English according to the text.

1. What does lean production refer to?

2. What principles need to be followed in lean production?

3. What are the basic elements of lean manufacturing? Which one is the most important?

4. What is the main impact of Kaizen?

Ⅲ. Read the text again and fill in the blanks in the following sentences orally.

1. Lean production is the fusion of various _____ philosophies designed to make operations as _____ as possible.

2. The philosophies invoked by lean production include _____ manufacturing, Kaizen, total _____ management(TQM), total _____ maintenance(TPM), _____ manufacturing(CM), etc.

3. Continuous improvement of processes throughout the value chain, mainly in terms of the _____ and _____.

4. To focus the production on customer orders means only products that are _____ _____ by customers should be produced.

5. _____ is a major influence on lean production.

IV. Translation.

Kaizen is a major influence on lean production. That's why lean manufacturing promotes teamwork among multi-skilled, multi-functional individuals at all levels, resulting in continuous process improvement of zero inventories, zero downtime, paperless, zero defects, and zero latency throughout the organization.

Lesson 29 Computer-integrated Manufacturing

Computer-integrated manufacturing (CIM) is a professional term used to describe modern manufacturing methods. CIM encompasses many other advanced manufacturing technologies such as computer numerical control (CNC), computer-aided design (CAD)/computer-aided manufacturing (CAM), robot technology, and just in time (JIT) delivery. Therefore, it is not only a new technology or a concept, computer-integrated manufacturing is also an entirely new manufacturing method and a business model.

In traditional manufacturing, only those processes that took place① in a workshop were considered manufacturing. Automation in machine tools in the workshop has not fundamentally changed the traditional thinking which divides the overall concept of manufacturing into separate specialized parts. Modern manufacturing includes all the activities and processes required to convert raw materials into② finished products, deliver them to the market, and support them on-site.

Margin Note

computer-integrated manufacturing
计算机集成制造

machine tool
机床

finished product
成品

These activities include:

(1) Identify the need for the product.

(2) Design products to meet requirements③.

(3) Obtain raw materials for production.

(4) Convert raw materials into finished products with appropriate processing technologies.

(5) Ship products to the market.

(6) Maintain products to ensure proper performance on spot.

This broad and modern view of manufacturing includes pre-production market analysis, research, development, and design, as well as post-production product delivery and product maintenance. With CIM, not only are the various processing cells automated but also the islands of automation are all linked or integrated. Integration means that the system can provide complete and instantaneous sharing of information. In modern manufacturing, integration is accomplished by computers. Therefore, computer-integrated manufacturing is the comprehensive integration of all components involved in converting raw materials into finished products and launching the products to the market.

processing cell
加工单元
island of automation
自动化孤岛

As shown in Fig. 29-1, computer-integrated manufacturing system (CIMS) consists of management, techniques, people, logistics, and information flow, while the development of computer technology has become an important support of advanced manufacturing technology, information technology, automation technology, and system engineering technology.

information flow
信息流

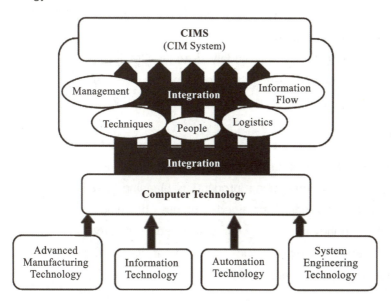

Fig. 29-1 CIMS & Computer Technology

With the development of computer technology, manufacturing has developed a complete cycle. The design has evolved from[④] a manual process with pencils, scales, and erasers into an automated computer-aided design (CAD) process. Process planning has evolved from a manual process with schedules, schematics, and diagrams to an automated process of computer-aided process planning (CAPP). Production has evolved from manual control machines to an automated process known as computer-aided manufacturing (CAM). They are

process planning
工艺规划
schematic
示意图

separate components that make up the design, planning, and production of CIM.

Material requirements planning (MRP) is an important concept that is directly related to CIM. It is a plan that can be used to calculate the number of raw materials that must be obtained to produce a specific quantity of a certain product. Material requirements planning involves the bill of materials, production schedules, and inventory records to make a comprehensive and detailed schedule for raw materials and components needed for a job.

bill of material
物料清单
inventory record
库存记录

As manufacturing technology has evolved from automation to integration, MRP has also evolved. MRP is now defined as[5] manufacturing resource planning. This broader concept refers to the determinization of material requirements. In addition, it includes financial tracking and accounting.

Lead time is important because it is used to make schedules for ordering materials and producing products. If lead time is not accurate, the MRP results will be equally inaccurate.

lead time
订货交付时间

An advanced manufacturing technology widely used in CIM is an automatic guided vehicle (AGV). With the development of the manufacturing industry to fully automated factories, AGVs will play an increasingly important role.

In conclusion, CIM refers to the integrated use of electronic data processing in all production-related enterprise departments. It includes information technology collaboration among production planning and control, computer-aided design, computer-aided process planning, computer-aided manufacturing, and computer-aided quality management, in which various technical and management functions required for the production should be integrated.

 New Words and Phrases

separate [ˈsepəreɪt] adj. 单独的,分开的,不同的,不相关的

accomplish [əˈkʌmplɪʃ] v. 完成,实现

schematic [skiːˈmætɪk] n. (尤指电子电路的)示意图

technique [tekˈniːk] n. 工艺,技巧

computer-integrated manufacturing 计算机集成制造
processing cell 加工单元
island of automation 自动化孤岛
information flow 信息流
process planning 工艺规划
inventory record 库存记录

 Notes

1. take place 发生,举行

例句:In traditional manufacturing, only those processes that took place in a workshop were considered manufacturing.

在传统的制造方式中,只有在车间进行的那些过程才被视为制造。

例句:It is envisaged that the talks will take place in the spring.

谈判预期在春季举行。

2. convert into 把……转换成……

例句:Modern manufacturing includes all the activities and processes required to convert raw materials into finished products, deliver them to the market, and support them on-site.

现代制造业包含了将原材料转化为成品,将其交付给市场,并在现场为其提供支持所需的所有活动和过程。

例句:It takes 15 minutes to convert the plane into a car by removing the wings and the tail.

拆除机翼和机尾,把那架飞机改成汽车需要15分钟的时间。

3. meet requirement 满足需求,达到要求

例句:Design products to meet requirements.
设计满足需求的产品。

例句:The commission said the gas producer must meet specific requirements before it can proceed.

委员会称,这家天然气生产商必须达到特定的要求才能继续运营。

4. evolve from… 由……进化

例句:The design has evolved from a manual process with pencils, scales, and erasers into an automated computer-aided design (CAD) process.

设计已经从使用铅笔、标尺和橡皮擦的手工过程演变为计算机辅助设计(CAD)的自动化过程。

例句:Birds are widely believed to have evolved from dinosaurs.
普遍认为鸟类是从恐龙进化而来的。

5. be defined as... 被定义为……，被称为……

例句：MRP is now defined as manufacturing resource planning.
MRP 现在的定义是制造资源计划。

例句：Procrastination is defined as the action of delaying or postponing something.
拖延症被定义为延迟或推迟某种事情的行为。

 Exercises

Ⅰ. **Write True or False beside the following statements about the text.**

1. _____ Computer-integrated manufacturing is a professional term used to describe modern manufacturing methods.

2. _____ Computer-integrated manufacturing is just a new manufacturing method.

3. _____ Automation in machine tools in the workshop has completely changed traditional thinking.

4. _____ In modern manufacturing, integration is accomplished by computers.

5. _____ Material requirements planning is indirectly related to computer-integrated manufacturing.

Ⅱ. **Answer the following questions in English according to the text.**

1. What does computer-integrated manufacturing refer to?
2. What activities are involved in modern manufacturing?
3. What is the current definition of MRP?
4. Why is lead time important?

Ⅲ. **Read the text again and fill in the blanks in the following sentences orally.**

1. Computer-integrated manufacturing encompasses many other advanced manufacturing technologies such as computer _____ control (CNC), computer-aided design (CAD)/_____ _____ (CAM), robot technology, and just in time (JIT) delivery.

2. With CIM, not only are the various _____ _____ automated but also the islands of automation are all linked or integrated.

3. Material requirements planning involves the _____ of _____, production schedules and inventory records to make a comprehensive and detailed schedule for raw materials and components needed for a job.

4. MRP is now defined as manufacturing _____ planning.

5. An advanced manufacturing technology widely used in CIM is an _____ _____ vehicle (AGV).

Ⅳ. **Translation.**

In conclusion, CIM refers to the integrated use of electronic data processing in all production-related enterprise departments. It includes information technology collaboration among production planning and control, computer-aided design, computer-aided process planning, computer-aided manufacturing, and computer-aided quality management, in which various technical and management functions required for the production should be integrated.

 ## Lesson 30　Collaborative Robots

A collaborative robot, as the name implies, is a robot intended for① direct human-robot interaction (HRI) within a shared space, or where human workers and robots are in close proximity②, giving full play to③ the efficiency of robots and human intelligence. Industrial robots have traditionally worked separately from humans behind fences or other protective barriers, but a collaborative robot, as a new type of industrial robot, removes the separation and interacts with human workers safely and directly in the designated area, as shown in Fig. 30-1.

Margin Note

collaborative robot
协作机器人

protective barrier
防护屏障

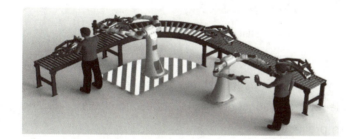

Fig. 30-1　**Collaborative Robots Collaborate with Humans on the Plant Floor**

Industrial collaborative robots help to automate unergonomic tasks, such as moving heavy parts, machine feeding, and assembly operations. Four levels of collaboration between industrial robots and human workers are defined as follows.

(1) Coexistence. Humans and robots work alongside each other without a fence. They do not, however, share a common workspace and work independently of one another on different tasks, as shown in Fig. 30-2.

(2) Sequential Collaboration. Humans and robots are active in the shared workspace, but their motions are sequential. They do not work on a part at the same time.

Fig. 30-2　Coexistence

(3) Cooperation. Robots and humans work on the same part at the same time, with both in motion④.

(4) Responsive Collaboration. The robot responds in real-time to the movements of the human worker, as shown in Fig. 30-3.

Fig. 30-3　Responsive Collaboration

In most of the industrial applications of collaborative robots today, the robot and the human worker share the same space but they complete tasks independently or sequentially. Cooperation and responsive collaboration are presently less common.

The great potential and benefits of human-robot collaboration are becoming increasingly evident in industrial communities that are influenced by a shift from mass production to highly customized and low-volume manufacturing processes.

human-robot collaboration
人机协作
mass production
大批量生产

Collaborative robots can automatize repetitive and high-effort tasks and reduce human task load by providing physical assistance, which may potentially improve the working conditions of human workers.

On the other hand, humans have better cognitive capabilities so that they can supervise robot operation or transfer new skills to the collaborative robot, thus adding a certain level of flexibility to the process and contributing to the effective accomplishment of a broad range of manufacturing tasks.

cognitive capability
认知能力

One of the most evident problems that arise from the integration of the human co-worker into the robot's workspace is human safety. Ensuring a safe interaction between the human and robot counterparts should be the main prerequisite of any collaborative robot control. The prominent examples of safety strategies to avoid physical contact between robots and humans are collision detection and reactive motion planning techniques. The concept of a safety map was recently introduced to give information about human injury occurrence and inherent global or task-dependent safety properties of a robot to the controller in a unified manner. Furthermore, some researchers have proposed to use of expert human demonstrations in an attempt to[5] achieve safe collaborative behavior of the robot.

collision detection
碰撞检测
reactive motion planning
反应式运动规划

The differences between collaborative robots and traditional industrial robots lie in[6] the following aspects.

(1) Ultra-small powerful servo drives mounted directly on the joints of collaborative robots are compact and can withstand extremely high mechanical acceleration and deceleration within the joints. The servo drives are closer to the encoder, which saves cable and reduces interference.

servo drive
伺服驱动器

(2) The double closed-loop control algorithm can improve the performance of each axis servo motor and the positioning accuracy of the endpoint.

double closed-loop control algorithm
双闭环控制算法
axis servo motor
轴伺服电机

(3) Motion redundancy. Robots with more than six axes are collectively referred to as redundant DOF (degree of freedom) robots. Compared with the traditional six-axis robot (as shown in Fig. 30-4), a seven-axis robot (as shown in Fig. 30-5) can extend the robot arm at multiple angles. Redundant DOF robots have more advantages in obstacle avoidance, overcoming singularities, flexibility, and fault tolerance.

redundant DOF robot
冗余自由度机器人

fault tolerance
容错性

(4) Sensors. A collaborative robot is equipped with cameras, video cameras, load sensors, laser safety sensors, and other sensing elements, which can sense the existence of human workers and take the corresponding action to avoid any harm to them. The part detection sensor can give feedback on the position of the gripper. If the gripper misses a part, the system will detect the error and repeat the operation once to ensure that the part is properly grabbed.

load sensor
负载传感器

part detection sensor
零件检测传感器

Fig. 30-4　Six-axis Robot

Fig. 30-5　Seven-axis Robot

Tips

1. DOF: Degree of Freedom 自由度

机器人机构能够独立运动的关节数目,称为机器人机构的运动自由度,简称自由度,简写为 DOF。机器人轴的数量决定了其自由度。其自由度越大越接近人手的动作机能,通用性就越好;但是自由度越大,结构越复杂,对机器人的整体要求就越高。目前在工业领域中,六轴机器人应用最为广泛。它与人类的手臂极为相似,其作用是移动末端执行器,在机械臂末端安装适用于特定应用场景的各种执行器,例如爪手、喷灯、钻头和喷漆器等,以完成不同的工作任务。

New Words and Phrases

collaborative [kəˈlæbərətɪv] *adj*. 协作的,合作的
collaborative robot 协作机器人
human-robot collaboration 人机协作
proximity [prɒkˈsɪməti] *n*. (时间或空间)接近,邻近,靠近
protective [prəˈtektɪv] *adj*. 防护的,保护的
protective barrier 防护屏障
designate [ˈdezɪɡneɪt] *v*. 指定,认定,委任,标明,(已当选)尚未就职
unergonomic [ˌʌnɜːɡəˈnɒmɪk] *adj*. 不适于提升工效的
coexistence [ˌkəʊɪɡˈzɪstəns] *n*. 共存,共处

alongside [əˌlɒŋˈsaɪd] prep. 与……一起，与……同时，在……旁边
sequential [sɪˈkwenʃl] adj. 顺序的，序列的，按次序的
evident [ˈevɪdənt] adj. 显然的，清楚的，显而易见的
mass production 大批量生产
cognitive [ˈkɒɡnətɪv] adj. 认知的，感知的，认识的
cognitive capability 认知能力
supervise [ˈsuːpəvaɪz] v. 监督，管理，指导，主管
counterpart [ˈkaʊntəpɑːt] n. 对应的事物，职位（或作用）相当的人
prerequisite [ˌpriːˈrekwəzɪt] n. 先决条件，前提，必备条件
prominent [ˈprɒmɪnənt] adj. 突出的，显眼的，显著的，凸显的
collision [kəˈlɪʒn] n. 碰撞（或相撞）事故，（意见、看法的）冲突，抵触
collision detection 碰撞检测
reactive motion planning 反应式运动规划
servo [ˈsɜːvəʊ] n. （机器的）伺服系统，随动系统
servo drive 伺服驱动器
double closed-loop control 双闭环控制算法
axis servo motor 轴伺服电机
redundant DOF robot 冗余自由度机器人
singularity [ˌsɪŋɡjuˈlærəti] n. 奇点，异常，奇特，奇怪
tolerance [ˈtɒlərəns] n. 容忍，忍受
fault tolerance 容错性
load sensor 负载传感器
part detection sensor 零件检测传感器

Notes

1. (be) intended for 为……打算（或设计）的

例句：A collaborative robot, as the name implies, is a robot <u>intended for</u> direct human-robot interaction (HRI) within a shared space, or where human workers and robots are in close proximity, giving full play to the efficiency of robots and human intelligence.

顾名思义，协作机器人是指在人机共享空间内，或者人类与机器人距离很近的情况下，实现人机直接交互，充分发挥机器人效率和人类智能的机器人。

例句：The books <u>were intended for</u> the edification of the masses.
这些书旨在教育民众。

2. in close proximity 紧邻，近在咫尺

例句：A collaborative robot, as the name implies, is a robot intended for direct human-robot interaction (HRI) within a shared space, or where human workers and robots are <u>in</u>

close proximity, giving full play to the efficiency of robots and human intelligence.

顾名思义，协作机器人是指在人机共享空间内，或者人类与机器人距离很近的情况下，实现人机直接交互，充分发挥机器人效率和人类智能的机器人。

例句：The same applies to colleagues who work in close proximity.

这同样适用于在附近工作的同事。

3. give full play to sth. 充分发挥，发扬

例句：A collaborative robot, as the name implies, is a robot intended for direct human-robot interaction (HRI) within a shared space, or where human workers and robots are in close proximity, giving full play to the efficiency of robots and human intelligence.

顾名思义，协作机器人是指在人机共享空间内，或者人类与机器人距离很近的情况下，实现人机直接交互，充分发挥机器人效率和人类智能的机器人。

例句：I need a job that can give full play to my skills.

我要找一份能充分发挥我技能的工作。

4. in motion 运转中，运动中

例句：Robots and humans work on the same part at the same time, with both in motion.

机器人和人类同时处理同一个部件，两者都在活动。

例句：The wheels of change have been set in motion.

变革的车轮已经开始运转。

5. in an attempt to do sth. 力图，试图，企图

例句：Furthermore, some researchers have proposed to use of expert human demonstrations in an attempt to achieve safe collaborative behavior of the robot.

此外，一些研究人员建议使用专家演示，以实现机器人与工作人员的安全协作。

例句：He came into the office in a vain attempt to have the report suppressed.

他走进办公室，试图压下这篇报告，但没有成功。

6. lie in （事情、问题）在于……

例句：The differences between collaborative robots and traditional industrial robots lie in the following aspects.

协作机器人和一般工业机器人的不同之处在于以下几个方面。

例句：The problem lies in deciding when to intervene.

问题在于决定何时介入。

 Exercises

Ⅰ. Write True or False beside the following statements about the text.

1. _____ Industrial collaborative robots help to automate unergonomic tasks, such

as moving heavy parts, machine feeding, and assembly operations.

2. _____ Cooperation and responsive collaboration are very popular.

3. _____ Industrial robots have better cognitive capabilities than humans.

4. _____ Collision detection and reactive motion planning techniques help to avoid physical contact between robots and human workers.

5. _____ Robots with less than six axes are collectively referred to as redundant DOF (degree of freedom) robots.

Ⅱ. Answer the following questions in English according to the text.

1. What is a collaborative robot?

2. Explain the four levels of collaboration between industrial robots and human workers.

3. What are the differences between collaborative robots and traditional industrial robots?

Ⅲ. Read the text again and fill in the blanks in the following sentences orally.

1. A collaborative robot, as a new type of _____ robot, removes the _____ and _____ with human workers _____ and _____ in the _____ area.

2. The great _____ and benefits of human-robot _____ are becoming increasingly _____ in industrial communities that are influenced by a shift from _____ production to highly _____ and _____ manufacturing processes.

3. Collaborative robots can _____ _____ and _____ tasks and reduce human _____ _____ by providing physical _____, which may potentially improve the working conditions of human workers.

4. One of the most evident problems that _____ _____ the _____ of the human co-worker into the robot's workspace is _____ _____. Ensuring a safe _____ between the human and robot _____ should be the main _____ of any _____ robot control.

5. The concept of a _____ _____ was recently introduced to give information about human _____ occurrence and _____ global or _____ safety properties of a robot to the _____ in a _____ manner. Furthermore, some researchers have _____ to use of expert _____ _____ in an _____ to achieve safe collaborative behavior of the robot.

Ⅳ. Translation.

Robots with more than six axes are collectively referred to as redundant DOF (degree of freedom) robots. Compared with the traditional six-axis robot, a seven-axis robot can extend the robot arm at multiple angles. Redundant DOF robots have more advantages in obstacle avoidance, overcoming singularities, flexibility, and fault tolerance.

Reading Material 21 Machine Learning

Machine learning (ML) is a field devoted to understanding and building learning methods, that is, leveraging data to improve performance on some sets of tasks. It is seen as a part of artificial intelligence. As shown in Fig. R21-1, machine learning algorithms build a model based on sample data, known as training data, to make predictions or decisions without being explicitly programmed to do so. Machine learning algorithms are used in a wide variety of applications, such as in intelligent manufacturing, intelligent medicine diagnosis and treatment, email filtering, speech recognition, and computer vision, where it is difficult or unfeasible to develop conventional algorithms to perform the needed tasks.

Margin Note

machine learning
机器学习

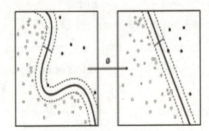

Fig. R21-1   Machine Learning

A subset of machine learning is closely related to computational statistics, which focuses on making predictions using computers, but not all machine learning is statistical learning. The study of mathematical optimization delivers methods, theories, and application fields for machine learning. Data mining is a related field of study, focusing on exploratory data analysis through unsupervised learning. Some machine learning is realized by using data and neural networks to mimic the way that the biological brain works. In cross-industry applications, machine learning is also known as predictive analytics.

computational statistic
计算机统计学

data mining
数据挖掘

predictive analytics
预测分析

Learning algorithms work on the basis that strategies, algorithms, and inferences worked well in the past and tend to continue working well in the future. These inferences can be obvious. For example, since the sun rose every morning for the last 10 000 days, it will probably rise tomorrow morning as well. They can be nuanced. For instance, $X\%$ of families have geographically separate species with color variants, so the probability of undiscovered black swans is Y%.

Machine learning programs can perform tasks without explicit programming. It involves computers learning from data provided so that they carry out certain tasks. For simple tasks assigned to computers, algorithms can be written to tell the machine how to execute all steps required to solve the problem at hand; for computers, there is a need to learn. For more advanced tasks, creating the required algorithms manually can be challenging. In practice, it can turn out to be more effective to help the machine develop its own algorithm, rather than having human programmers specify every needed step.

The discipline of machine learning employs various approaches to teach computers to accomplish tasks where no fully satisfactory algorithm is available. Where there are a lot of potential answers, one approach to do this is to label some of the correct ones as valid. This can then be used as training data for the computer to improve its algorithm to determine the correct answers. For example, to train a system for the task of digital character recognition, the MNIST (mixed national institute of standards and technology database) dataset of handwritten digits has often been used.

New Words and Phrases

leverage [ˈliːvərɪdʒ] v. 充分利用（资源、观点等） n. 影响力，手段，杠杆作用
explicitly [ɪkˈsplɪsɪtli] adv. 清楚明确地，详述地
exploratory [ɪkˈsplɒrət(ə)ri] adj. 探究的，勘探的，考察的
unsupervised [ˌʌnˈsuːpəvaɪzd] adj. 无人监督的，无人管理的
neural [ˈnjʊərəl] adj. 神经的，神经系统的
nuanced [ˈnjuːɒnst] adj. 微妙的，具有细微差别的
geographically [ˌdʒiːəˈɡræfɪkli] adv. 在地理上，地理学上
species [ˈspiːʃiːz] n. 种类，物种
swan [swɒn] v. 无目的地漫游，闲逛
discipline [ˈdɪsəplɪn] n. （尤指大学的）科目，学科，纪律，自制力
machine learning 机器学习
computational statistic 计算机统计学

data mining 数据挖掘
predictive analytics 预测分析

Reading Material 22   Digital Image Processing

Digital image processing generally refers to the processing of 2D pictures by a digital computer (as shown in Fig. R22-1). In a broader sense, it implies digital processing of any 2D data. A digital image is an array of real or complex numbers represented by a finite number of bits. An image given in the form of a transparency, slide, photograph, or chart is first digitized and stored as a matrix of binary digits in computer memory, the high-capacity RAM. This digitized image can then be processed and/or displayed on a high-resolution monitor which refreshes at 3D frames to produce a visibly continuous display. Small or microcomputers are used to communicate and control all the digitization, storage, processing, and display operations via a computer network (such as the ethernet). Program inputs to the computer are made through a terminal, and the outputs are available on a terminal, television monitor, or printer.

Margin Note

finite number
有限数

binary digit
二进制数字

ethernet
以太网

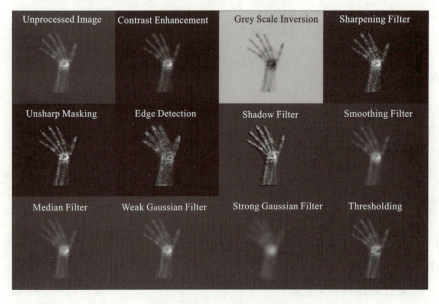

Fig. R22-1   Digital Image Processing

Digital image processing has a broad spectrum of applications, such as remote sensing via satellites and other spacecraft, image

remote sensing
遥感

transmission and storage for business applications, medicine, radar, sonar, acoustic image processing, robotics, and automated inspection of industrial parts.

    Images acquired by satellites are useful in the tracking of earth resources, geographical mapping, prediction of agricultural crops, urban growth and weather, flood and fire control, and many other environmental applications. Space image applications include recognition and analysis of objects contained in images obtained from deep space probe missions. Image transmission and storage applications occur in broadcast television, teleconferencing, the transmission of facsimile images (printed documents and graphics) for office automation, computer network communication, security monitoring systems based on closed-circuit television, and military communications. In medical applications, people are concerned with the processing of projection images from chest X-rays, cine-angiograms, transaxial tomography, and other medical images in radiology, nuclear magnetic resonance (NMR), and ultrasonic scanning. These images may be used for patient screening and monitoring or for the detection of tumors or other diseases in patients. Radar and sonar images are used for the detection and recognition of various types of targets or in the guidance and maneuvering of aircraft or missile systems. There are many other applications ranging from robot vision for industrial automation to image synthesis for cartoon making or fashion design. Whenever a human being or a machine or any other entity receives data of two or more dimensions, an image is processed.

acoustic image processing 声像处理

deep space probe mission 深空探测任务

closed-circuit television 闭路电视

transaxial tomography 轴向体层成像

nuclear magnetic resonance 核磁共振

ultrasonic scanning 超声波扫描

 New Words and Phrases

dimensional [daɪˈmenʃən(ə)l] *adj*. 维度的
array [əˈreɪ] *n*. 一系列,大量
matrix [ˈmeɪtrɪks] *n*. 矩阵,模型
high-resolution 高清晰度的,高分辨率的
sonar [ˈsəʊnɑː(r)] *n*. 声呐,声波定位仪
facsimile [fækˈsɪməli] *n*. 传真,复写
projection [prəˈdʒekʃn] *n*. 投射,投影

tumor ['tjuːmə(r)] n. 肿瘤,肿块
guidance ['gaɪdns] n. (火箭等的)制导,指导,指引
synthesis ['sɪnθəsɪs] n. 合成,综合,综合体
finite number 有限数
binary digit 二进制数字
remote sensing 遥感
acoustic image processing 声像处理
deep space probe mission 深空探测任务
closed-circuit television 闭路电视
transaxial tomography 轴向体层成像
nuclear magnetic resonance 核磁共振
ultrasonic scanning 超声波扫描

Reading Material 23 The Latest Intelligent Manufacturing Technology

Intelligent manufacturing is one of the latest manufacturing models, with broad prospects for development, intelligent manufacturing is essentially an intelligent information processing system, external control of robot actions, and complete the manufacturing and processing of products.

Margin Note

1. Cyber-physical System

Cyber-physical system (CPS) is a multidimensional complex system that integrates computing, network, and physical environment. As shown in Fig. R23-1, through the organic integration and deep collaboration of 3C (computing, communication, control) technologies, it can realize real-time perception, dynamic control and information service of a large-scale engineering system, and endow physical equipment with five functions: computing, communication, precise control, remote coordination, and autonomy, so as to achieve the integration of virtual network world and real physical world.

2. Artificial Intelligence

Artificial intelligence (AI) is the theory, method, technology, and application system to research and develop for simulating, extending,

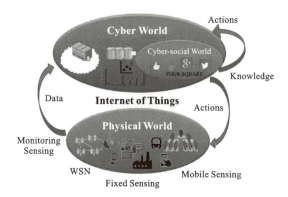

Fig. R23-1   Cyber-physical system

and expanding human intelligence. As shown in Fig. R23-2, it seeks to understand the nature of intelligence and produce a new kind of intelligent machine that can respond in a manner similar to human intelligence. The research in this field includes robotics, speech recognition, image recognition, natural language processing, and expert systems.

Fig. R23-2   Artificial Intelligence

3. Augmented Reality

Augmented reality (AR) technology is a new technology that "seamlessly" integrates real-world information and virtual-world information. Physical information (visual, sound, taste, tactile, and other information) is difficult to experience in a certain time and space in the real world. However, through computers and other science and technology, virtual information is applied to the real world after simulation and superimposition, so that the information can be perceived by human senses, and then achieve a sensory experience beyond reality. Augmented reality technology includes a range of new technologies and new means such as multimedia, 3D modeling, real-time video display and control, multi-sensor fusion, real-time tracking and registration, and scene fusion.

4. Model-based Enterprise

Model-based enterprise (MBE) is a manufacturing entity. It adopts modeling and simulation technology to completely improve, seamlessly integrate, and manage all technical and business processes of its design, manufacturing, and product support. It defines, executes, controls, and manages all processes of the enterprise on the basis of the product and process models. It uses scientific simulation and analysis tools to make the best decisions at every step of product lifecycle management (PLM), radically reducing the time and cost of product innovation, development, manufacturing, and support.

model-based enterprise
基于模型的企业

simulation technology
仿真技术

5. Internet of Things

Internet of things (IoT) is shown in Fig. R23-3. It refers to a huge network formed by the combination of the internet through various information sensing devices to collect real-time information required such as objects or processes that need monitoring, connection, and interaction. It aims at realizing the connection between things and things, things and people, all things and the network, which is convenient for identification, management, and control.

internet of things
物联网

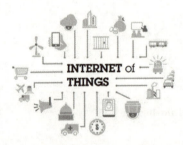

Fig. R23-3　Internet of Things

6. Cloud Computing

Cloud computing (CC) is a payment pattern in terms of actual usage. The pattern provides available, convenient, on-demand network access to the configurable computing shared resources pool (resources including network, servers, storage, applications, and services). These resources can be provided quickly with little management work or minimal interaction with service providers.

cloud computing
云计算

7. Industrial Big Data

Industrial big data (IBD) refers to the application of the big data concept in the industrial field. It realizes the interconnection of originally

industrial big data
工业大数据

isolated, massive, and diverse data such as equipment data, activity data, environment data, service data, business data, market data, and upstream and downstream industrial chain data. In this way, it achieves the connection between people, things and things, people and things, especially the connection between end users and manufacturing as well as service processes. At the same time, it realizes the mutual conversion among data, information, and knowledge, so that it has stronger decision-making power, insight and discovery power, and process optimization ability.

8. Prognostics and Health Management

Prognostics and health management (PHM) is a new approach to health management based on modern information technology and artificial intelligence. PHM system mainly consists of six parts: data collection, information induction and processing, condition monitoring, health assessment, fault prediction and decision-making, and guarantee decision.

prognostics and health management 故障预测与健康管理

9. Mixed Manufacturing

3D printing (additive manufacturing) technology and milling (subtraction manufacturing) technology are organically combined to form a new manufacturing mode. Hybrid manufacturing can effectively realize the machining of new geometric shapes by virtue of the advantages of additive manufacturing. Meanwhile, additive manufacturing technology is no longer limited to processing small workpieces, and the processing efficiency is also greatly improved.

mixed manufacturing 混合制造
additive manufacturing 增材制造
subtraction manufacturing 减材制造
hybrid manufacturing 混合制造

10. Factory Information Security

The terminal devices and systems involved in factory information security include industrial ethernet, data acquisition and monitoring system, distributed control system (DCS), process control system (PCS), programmable logic controller (PLC), remote monitoring system, and operation security of industrial control system. In this way, it ensures that industrial ethernet and industrial systems are not accessed, used, leaked, interrupted, modified, and destroyed without authorization, thus providing information services for the normal production of enterprises and the normal use of products.

New Words and Phrases

prospect ['prɒspekt] n. 前景,可能性,希望,预期,展望
large-scale [ˌlɑːdʒ'skeɪl] adj. 大规模的,大范围的
superimposition [ˌsuːpərˌɪmpə'zɪʃn] n. 添上,[摄]叠印,重叠
sensory ['sensəri] adj. 感觉的,感官的
scene [siːn] n. 事件,场面,事发地,现场
radically ['rædɪkli] adv. 根本上,彻底地
isolated ['aɪsəleɪtɪd] adj. 单独的,孤寂的,遥远的,偏僻的
diverse [daɪ'vɜːs] adj. 不同的,各式各样的
upstream [ˌʌp'striːm] adj. 向(在)上游的,逆流而上的
downstream [ˌdaʊn'striːm] adj. 顺流的,在(或向)下游的
mutual ['mjuːtʃuəl] adj. 相互的,彼此的,共同的,共有的
induction [ɪn'dʌkʃn] n. 归纳,归纳法,引导
guarantee [ˌɡærən'tiː] n. 保证,担保
subtraction [səb'trækʃn] n. 减去,(数)减,减法
geometric [ˌdʒiːə'metrɪk] adj. 几何图形的,几何的
by virtue of 由于,凭借
authorization [ˌɔːθəraɪ'zeɪʃn] n. 批准,授权,批准书,授权书
model-based enterprise 基于模型的企业
simulation technology 仿真技术
industrial big data 工业大数据
prognostics and health management 故障预测与健康管理
mixed manufacturing 混合制造
hybrid manufacturing 混合制造

 ## Reading Material 24 Intelligent Manufacturing Led by New Generation of AI Technology

The connotation of intelligent manufacturing under the guidance of the new generation of artificial intelligence technology is as follows in terms of the technical means, characteristics, implementation contents, modes, business formats, and objectives.

(1) Technical means. Intelligent manufacturing is based on the

Margin Note

ubiquitous network, with the help of digital, networked (interconnected), and intelligent technical tools which are deeply integrated with emerging manufacturing science and technology, information science and technology, intelligent science and technology, and technologies in the field of manufacturing applications. It constitutes a user-centered, unified service cloud (internet services) of wisdom manufacturing resources, products, and capabilities, enabling users to obtain wisdom manufacturing resources, products, and capabilities on demand anytime, anywhere through intelligent terminals and intelligent cloud manufacturing service platforms and further complete the activities of the entire manufacturing life cycle with high quality.

ubiquitous network
泛在网

wisdom manufacturing
智慧制造

(2) Characteristics. People, machines, objects, environment, and information in the whole manufacturing system and life cycle activities (industrial chain) are perceived, connected, collaborated, learned, analyzed, recognized, decided, controlled, and executed independently and intelligently.

(3) Implementation contents. Promote the integration and optimization of people, technology and equipment, management, data, materials, capital (six elements) and people flow, technology flow, management flow, data flow, logistics, capital flow (six streams) in the whole manufacturing system and life cycle activities.

(4) Modes. User-centered mode integrating people, machines, objects, environment, and information, and a new model of intelligent manufacturing which is interconnected (collaborated), service-oriented, personalized (customized), flexible, socialized, and intelligent.

(5) Business formats. Include ubiquitous connectivity, data-driven, shared services, cross-border integration, independent wisdom, and mass innovation.

(6) Objectives. To achieve efficient, high-quality, economical, green, and flexible manufacturing of products and services for users, and improve the market competitiveness of enterprises (or groups).

The architecture of an intelligent manufacturing system under the guidance of the new generation of artificial intelligence technology mainly includes five layers: new intelligent resources, capabilities, and products; new intelligent perception, access, and communication; new intelligent manufacturing cloud service platform; new intelligent manufacturing cloud service application; newcomers and organization.

The technical system of intelligent manufacturing system led by the new generation of artificial intelligence technology mainly includes the overall technology of intelligent manufacturing, platform technology, and intelligent technology of manufacturing life cycle activities.

It is also the industrial internet system, that is, the industrial cloud network, which is produced according to the national conditions and the development trend of the industrial industry in the information age.

 New Words and Phrases

connotation [ˌkɒnəˈteɪʃ(ə)n] n. 内涵意义,隐含意义
perceive [pəˈsiːv] vt. 察觉,注意到,认为,理解
capital [ˈkæpɪt(ə)l] n. 资本,资金
ubiquitous network 泛在网
wisdom manufacturing 智慧制造

Appendix Ⅰ　Vocabulary

A

	课文编号
abbreviate [ə'briːvieɪt] v. 缩略，缩写	R12
aberration [ˌæbə'reɪʃn] n. 脱离常规，反常现象，异常行为	R4
abnormal [æb'nɔːml] adj. 不正常的，反常的	10
abort [ə'bɔːt] v. 中止，中辍（计划、活动等）	15
absolute ['æbsəluːt] adj. 绝对的，无疑的	25
absolute value instruction 绝对值指令	25
acceleration [əkˌselə'reɪʃn] n. 加速，加快	6
accelerometer [əkˌselə'rɒmɪtə(r)] n. 加速度计，加速器	21
acceptance decision 验收决定	R15
access ['ækses] v. 访问，存取（计算机文件），到达，进入，使用	R16
accessory [ək'sesəri] n. 附件，配件，附属物	4
accidentally [ˌæksɪ'dentəli] adv. 意外地，偶然地	10
accomplish [ə'kʌmplɪʃ] v. 完成，实现	29
accounts payable 应付账款	R14
accounts receivable 应收账款	R14
accumulate [ə'kjuːmjəleɪt] v. 积累，积聚	3
accuracy ['ækjərəsi] n. 精确程度，准确性	R14
accurate ['ækjərət] adj. 精确的，准确的	3
acknowledgment [ək'nɒlɪdʒmənt] n. 承认，致谢	R10
acoustic image processing 声像处理	R22
acquisition [ˌækwɪ'zɪʃn] n. （知识、技能等的）获得，得到	R2
activate ['æktɪveɪt] v. 激活	4
active ['æktɪv] adj. 有源的，活跃的	4
active tag 有源标签	4
actuator ['æktjʊeɪtə] n. 执行机构（元件）	6
adaptive [ə'dæptɪv] adj. 适应的，有适应能力的	26
adaptive echo cancellation 自适应回声抵消	R10
additional cabinet 附加机柜	R7
additive ['ædətɪv] adj. 附加的，加成的，加法的	R17

词条	页码
add-on utility 附加实用程序	16
adequate [ˈædɪkwət] adj. 充足的,适当的,胜任的	22
adjacent [əˈdʒeɪsnt] adj. 相邻的,邻近的,与……毗连的	R12
adjust [əˈdʒʌst] v. 调整,调节	10
advanced [ədˈvɑːnst] adj. 先进的,高级的,高等的	16
advanced artificial intelligence system 高级人工智能系统	R10
aerospace [ˈeərəʊspeɪs] n. 航空航天工业/技术 adj. 航空和航天(工业)的	R19
agile [ˈædʒaɪl] adj. 灵活的,敏捷的,机敏的,机灵的	11
agile manufacturing environment 敏捷制造环境	18
agile manufacturing 敏捷制造	26
agility [əˈdʒɪlɪti] n. 敏捷性,灵敏性	1
airborne [ˈeəbɔːn] adj. 在空中的,飞行中的	3
airborne sensor 机载遥感器	3
alarm information 报警信息	20
algorithm [ˈælɡərɪðəm] n. 算法,运算法则	5
alignment [əˈlaɪnmənt] n. 对齐,排成直线	R13
alliance [əˈlaɪəns] n. 联盟,结盟,(人们之间的)亲密关系,(尤指)联姻	26
allowable [əˈlaʊəbl] adj. 允许的,承认的,容许的	10
alloying [əˈlɔɪɪŋ] n. 合金化处理,炼制合金	22
alloying element 合金元素	22
alongside [əˌlɒŋˈsaɪd] prep. 与……一起,与……同时,在……旁边	30
alter [ˈɔːltə(r)] v. (使)改变,更改,改动	R2
alternative [ɔːlˈtɜːnətɪv] adj. 可供替代的,另类的,非传统的 n. 可供选择的事物	3
aluminum [əˈluːmɪnəm] n. 铝	25
analog [ˈænəlɒɡ] adj. 模拟的,指针式的	R12
analog-to-digital converter (ADC) 模数转换	R12
analyze [ˈænəlaɪz] v. 分析(研究),分解,解析	3
anchor [ˈæŋkə(r)] v. 使固定	R5
anchor bolt 地脚螺栓	R5
angle [ˈæŋɡ(ə)l] n. 倾斜,斜角,角度	25
animation [ˌænɪˈmeɪʃn] n. 动画,动画片,活力	R10
anomaly [əˈnɒməli] n. 异常,异常事物,反常现象	3
anticlockwise [ˌæntiˈklɒkwaɪz] adj. 逆时针的	25
application [ˌæplɪˈkeɪʃn] n. (尤指理论、发现等的)应用,运用	1

application executive 应用执行器	R2
approach [əˈprəutʃ] n. 方法,态度,靠近,接近	19
appropriate [əˈprəupriət] adj. 合适的,相称的	26
approved vendor 认可供应商	R13
approximately [əˈprɒksɪmətli] adv. 大概,大约,约莫	14
arc [ɑːk] n. 弧形,弧线	25
architecture [ˈɑːkɪtektʃə(r)] n. 体系结构,(总体、层次)结构	R3
archives [ˈɑːkaivz] n. 档案,案卷(archive 的复数)	16
arm assembly 手臂组件	7
array [əˈreɪ] n. 一系列,大量	R22
articulated [ɑːˈtɪkjuleɪtɪd] adj. 有关节的,铰接的	6
articulated robot 多关节型机器人	6
artificial [ˌɑːtɪˈfɪʃl] adj. 人工的,人造的	1
artificial intelligence 人工智能	1
artificial neural network 人工神经网络	R1
as-build record 竣工记录	17
assembly [əˈsembli] n. 组装,装配	3
assembly line 装配线	3
assertion [əˈsɜːʃn] n. 主张,声称	R2
assume [əˈsjuːm] v. 承担,就职,假定,假设	27
assurance [əˈʃuərəns] n. 保证,确保	13
attach [əˈtætʃ] v. 贴上,把……固定,附上	4
audio [ˈɔːdiəu] n. 音频,音响	R10
audit [ˈɔːdɪt] vi , n. 审计,查账	R15
auditory [ˈɔːdətri] adj. 听觉的,听的	9
augmented [ɔːgˈmentɪd] adj. 增广的,增强的	3
augmented reality 增强现实	3
authentication [ɔːˌθentɪˈkeɪʃn] n. 身份验证,认证,鉴定	4
authorization [ˌɔːθəraɪˈzeɪʃn] n. 批准,授权,批准书,授权书	R23
automatic guided vehicle 自动导引车	R20
automatic identification and data capture 自动识别和数据捕获	4
automatic production line 自动生产线	20
automatic programming 自动编程	25
automotive [ˌɔːtəˈməutɪv] adj. 汽车的,自动的	2
autonomous [ɔːˈtɒnəməs] adj. 自主的,有自主权的	1

autonomous mobile robot 自主移动机器人　　　　　　　　　　　　　　　　　9

autonomy [ɔːˈtɒnəmi] n. 自主化，自主，自主权　　　　　　　　　　　　　12

available [əˈveɪləbl] adj. 可获得的，可找到的，有空的，未婚的　　　　　　15

average [ˈævərɪdʒ] v. 平均为 adj. 平均的 n. 平均数　　　　　　　　　　6

axis [ˈæksɪs] n. 轴（旋转物体假想的中心线）　　　　　　　　　　　　　　6

axis servo motor 轴伺服电机　　　　　　　　　　　　　　　　　　　　　30

B

backup battery 备用电池　　　　　　　　　　　　　　　　　　　　　　　8

balance [ˈbæləns] n. 平衡，均衡，均势　　　　　　　　　　　　　　　　7

balance drive bearing 平衡传动轴承　　　　　　　　　　　　　　　　　　7

balance spring rod 平衡弹簧杆　　　　　　　　　　　　　　　　　　　　7

barcode [ˈbɑːkəʊd] n. 条形码　　　　　　　　　　　　　　　　　　　　4

base metal area 母材区　　　　　　　　　　　　　　　　　　　　　　　19

baseplate [beɪspleɪt] n. 底板，撑板，底盘　　　　　　　　　　　　　　R5

batch processing 批处理　　　　　　　　　　　　　　　　　　　　　　　28

battery [ˈbætri] n. 电池　　　　　　　　　　　　　　　　　　　　　　4

battery clamp 电池夹　　　　　　　　　　　　　　　　　　　　　　　　8

battery connector 电池连接器　　　　　　　　　　　　　　　　　　　　8

bearing [ˈbeərɪŋ] n. 轴承，定向，方位　　　　　　　　　　　　　　　　7

benchmark [ˈbentʃmɑːk] n. 基准　　　　　　　　　　　　　　　　　　R1

bevel [ˈbevl] n. 斜角，斜面，[测]斜角规　　　　　　　　　　　　　　　7

beveled edge 斜边　　　　　　　　　　　　　　　　　　　　　　　　　25

beverage [ˈbevərɪdʒ] n. 饮料　　　　　　　　　　　　　　　　　　　　R8

bill of material 物料清单　　　　　　　　　　　　　　　　　　　　　　17

bin [bɪn] n. 仓，箱子　　　　　　　　　　　　　　　　　　　　　　　20

bin inventory 料仓盘点　　　　　　　　　　　　　　　　　　　　　　　20

bin status monitoring 料仓状态监控　　　　　　　　　　　　　　　　　20

binary [ˈbaɪnəri] adj. 二进制的（用0和1记数），二元的，由两部分组成的　4

binary digit 二进制数字　　　　　　　　　　　　　　　　　　　　　　R22

binary number 二进制数　　　　　　　　　　　　　　　　　　　　　　R4

binder [ˈbaɪndə] n. 黏结剂　　　　　　　　　　　　　　　　　　　　24

biodiversity [ˌbaɪəʊdaɪˈvɜːsiti] n. 生物多样性　　　　　　　　　　　27

bionic [baɪˈɒnɪk] adj.（因体内有电子装置）能力超人的　　　　　　　　6

bit [bɪt] n. 比特　　　　　　　　　　　　　　　　　　　　　　　　　R4

blank [ˈblæŋk] n. 坯料，空白处，空格 adj. 单调的，彻底的　　　　　　18

blind [blaɪnd] n. 窗帘,用以蒙蔽人的言行,借口	R9
block diagram 框图	R12
bolt [bəʊlt] n. 螺栓	R5
bombard [bɒmˈbɑːd] v. 冲击,轰击	21
boost [buːst] v. 使增长,推动	27
botch [bɒtʃ] v. 笨拙地弄糟	R4
bottom line 最终盈利(或亏损)	27
boundary [ˈbaʊndri] n. 边界,界限,分界线	R4
boundary layer 边界层	21
brake [breɪk] n. 刹车,车闸	R7
branch [brɑːntʃ] n. 分支,树枝,分部	R10
bridge [brɪdʒ] v. 弥合,桥梁	R9
brightness distribution 亮度分布	19
broadly [ˈbrɔːdli] adv. 广泛地,普遍地,大体上	27
bulk micromachining 体微加工	21
burr-free 无毛刺	R19
button [ˈbʌt(ə)n] n. 按钮,纽扣,扣子	R7
by virtue of 由于,凭借	R23

C

cabinet [ˈkæbɪnət] n. 机箱,储藏柜,陈列柜	8
calculate [ˈkælkjuleɪt] v. 计算,核算,预测,推测	16
calibration [ˌkælɪˈbreɪʃn] n. 标定,(测量器上的)刻度	8
calibration mark 校准标记	8
cantilever gate electrode 悬臂栅电极	21
capacity requirement 能力要求	R14
capital [ˈkæpɪt(ə)l] n. 资本,资金	R24
capture [ˈkæptʃə(r)] v. 捕获	4
carbon dioxide (CO_2) 二氧化碳	R7
carbon steel 碳素钢	22
cart [kɑːt] n. 手推车,手拉车	4
Cartesian [kɑːˈtiːziən] 笛卡儿	15
Cartesian robot 直角坐标型机器人	6
carton [ˈkɑːtn] n. 硬纸盒,塑料盒	R8
casting [ˈkɑːstɪŋ] n. 铸件,铸造物,角色分配,演员挑选	19
catastrophic [ˌkætəˈstrɒfɪk] adj. 灾难性的	11

category [ˈkætɪg(ə)rɪ] n. 种类,分类,类别	R15
cavity [ˈkævɪtɪ] n. 腔,洞,凹处	22
cell [sel] n. 小隔间,细胞	R7
cellular [ˈseljələ(r)] adj. (无线电话)蜂窝状的,细胞的,网状的	11
cellular manufacturing 单元式制造	28
cement [sɪˈment] n. 水泥	R8
ceramic [sɪˈræmɪk] adj. 陶瓷的	21
charge [tʃɑːdʒ] v. 充电,使……承担责任	8
charge coupled device 电荷耦合器件	19
chart [tʃɑːt] n. 图表	R15
checklist [ˈtʃeklɪst] n. 检查表	R15
chemical etching 化学腐蚀法	21
chip [tʃɪp] n. 碎块,碎屑,芯片	R19
circuit [ˈsɜːkɪt] n. 线路,电路,环形路线	14
circuit board 电路板	19
circular [ˈsɜːkjələ(r)] adj. 循环的,圆形的,环形的	27
circular economy 循环经济	27
circulate [ˈsɜːkjuleɪt] v. 循环,(使)环行,环流	23
clamp [klæmp] n. 夹具,钳位电路	8
classify [ˈklæsɪfaɪ] v. 分类,划分,将……分类	4
clip [klɪp] n. 夹子,别针,片段	R8
clip-and-grab mechanical gripper 夹抓式机械抓手	R8
clockwise [ˈklɒkwaɪz] adj. 顺时针的	25
clockwise/anticlockwise arc interpolation 顺时针/逆时针圆弧插补	25
closed-circuit television 闭路电视	R22
closed-loop optimization 闭环优化	12
cloud computing 云计算	12
CNC program 计算机数控程序	16
coarse-acquisition (C/A) code 粗捕获码	R11
coating [ˈkəʊtɪŋ] n. 涂料,涂层	R8
code division multiple access (CDMA) 码分多址	R11
coexistence [ˌkəʊɪgˈzɪstəns] n. 共存,共处	30
cognitive [ˈkɒgnətɪv] adj. 认知的,感知的,认识的	30
cognitive capability 认知能力	30
collaborative [kəˈlæbərətɪv] adj. 协作的,合作的	30

collaborative robot 协作机器人	30
collision [kəˈlɪʒn] n. 碰撞(或相撞)事故,(意见、看法的)冲突,抵触	30
collision detection 碰撞检测	30
combined mechanical gripper 组合式机械抓手	R8
command [kəˈmɑːnd] n. (计算机的)指令,命令,指示	25
commitment [kəˈmɪtmənt] n. 承诺,保证,奉献,投入	27
communication [kəmjuːnɪˈkeɪʃ(ə)n] n. 通讯,通信,交通,联络	5
compact [ˈkɒmpækt] adj. 紧凑的,紧密的,体积小的	6
compatible [kəmˈpætəbl] adj. 可共用的,兼容的	3
compensation [ˌkɒmpenˈseɪʃ(ə)n] n. 弥补,抵消,赔偿金,补偿金	25
competitive [kəmˈpetətɪv] adj. 竞争的	26
compilation [ˌkɒmpɪˈleɪʃn] n. 汇编,编写	R16
compile [kəmˈpaɪl] v. 编写,编纂	R12
compiler [kəmˈpaɪlə(r)] n. 编纂者,汇编者	R12
complementary [ˌkɒmplɪˈmentri] adj. 互补的,补充的,相互补足的	3
component [kəmˈpəʊnənt] n. 组成部分,成分,部件	1
comprehensive [ˌkɒmprɪˈhensɪv] adj. 综合的,全面的	2
compressed [kəmˈprest] adj. (被)压缩的,扁的	8
compressed air 压缩空气	8
computational statistic 计算机统计学	R21
computer-aided design 计算机辅助设计	3
computer-integrated manufacturing 计算机集成制造	29
computer-aided process planning 计算机辅助工艺设计	18
concatenate [kɒnˈkæt(ə)ˌneɪt] vt. 连接,拼接 adj. 连锁的	1
conception [kənˈsepʃn] n. 概念,观念,构想,设想	1
concurrent [kənˈkʌrənt] adj. 同时发生的,并存的	R12
concurrent engineering 并行工程	R13
concurrent operation 并行操作系统	R12
conducive [kənˈdjuːsɪv] adj. 有助于(有利于)……的	13
conducting electricity 感应电流	23
conducting [ˈkɒndʌktɪŋ] adj. 传导的 n. 指挥,传导	23
configuration [kənˌfɪɡəˈreɪʃn] n. 配置,结构,构造	20
configuration management 配置管理	R13
configure [kənˈfɪɡə(r)] v. (计算机)配置,安装,设定	16
confirm [kənˈfɜːm] v. 确定,确认,证实,证明,批准,认可,使坚定,加强	15

conflict [ˈkɒnflɪkt] v. 冲突，抵触   2
conform [kənˈfɔːm] vi. 符合，遵照，适应环境   R15
connotation [ˌkɒnəˈteɪʃ(ə)n] n. 内涵意义，隐含意义   R24
consequence [ˈkɒnsɪkwəns] n. 结果，后果   R2
consistent [kənˈsɪstənt] adj. 一致的，坚持的，坚固的   5
consistent with 与某事物并存（一致）   R5
consolidate [kənˈsɔliˈdeit] vt. 巩固，合并，统一   24
consolidation [kənˌsɒlɪˈdeɪʃn] n. 整合，巩固   R16
constant value 恒定值   23
constantly [ˈkɒnstəntli] adv. 始终，一直，重复不断地   1
constitute [ˈkɒnstɪtjuːt] v. 组成，构成   1
constraint [kənˈstreɪnt] n. 限制，限定，约束   3
consult [kənˈsʌlt] v. 咨询，请教，商量，商讨   R7
consultation [ˌkɒnslˈteɪʃn] n. 咨询，商讨，磋商，协商会   13
consumption [kənˈsʌmpʃn] n. 消耗   3
contaminated [kənˈtæməneɪtɪd] adj. 受污染的   7
contention [kənˈtenʃn] n. 争用，争吵，争论，看法，观点   11
continuous [kənˈtɪnjuəs] adj. 连续不断的，持续的   28
continuous product flow 连续产品流   28
continuum [kənˈtɪnjuəm] (pl. continua) n. 连续体   28
contour [ˈkɒntʊə(r)] n. 轮廓，外形，周线   25
control algorithm 控制算法   5
control cabinet 控制柜   8
control chart 控制图   R15
control module 控制模块   R7
conventional [kənˈvenʃnl] adj. 传统的，习惯的   R2
conversion [kənˈvɜːʃn] n. 转换，转变，改变，归附   19
convert [kənˈvɜːt] v. 转换，(使)转变，转化，可转变为，可变换成   23
converter [kənˈvɜːtə] n. 转换器，使发生转化的人(或物)   R12
conveyor [kənˈveɪə(r)] n. 传输带，运送者，传播者   9
conveyor belt 传送带   9
coolant [ˈkuːlənt] n. 冷却剂   16
cooperation [kəʊˌɒpəˈreɪʃn] n. 合作，协作   26
cooperative [kəʊˈɒpərətɪv] adj. 协作的，同心协力的，配合的   9
coordinate [kəʊˈɔːdɪneɪt] n. 坐标, v. 协调，配合   25

coordinate rotation 坐标轴旋转	25
coordination [kəʊˌɔːdɪˈneɪʃn] n. 协作, 协调, 配合	1
corresponding [ˌkɒrəˈspɒndɪŋ] adj. 相应的, 符合的, 相关的	13
corrosion [kəˈrəʊʒ(ə)n] n. 腐蚀, 侵蚀	22
corrosion resistance 耐蚀性	22
cost-effectively 有成本效益地, 划算地	2
counterfeit [ˈkaʊntəfɪt] v. 伪造, 仿造, 制假	R9
countermeasure [ˈkaʊntəmeʒə(r)] n. 对策, 对抗手段, 反措施	R5
counterpart [ˈkaʊntəpɑːt] n. 对应的事物, 职位(或作用)相当的人	30
coupled [ˈkʌp(ə)ld] adj. 耦合的, 联结的	19
CPU 中央处理器	R12
crack [kræk] n. 缝隙, 狭缝, 窄缝	R4
crawl [krɔːl] v. 爬行, 匍匐前进, 缓慢前进	9
crawling robot 爬行机器人	9
critical [ˈkrɪtɪkl] adj. 极其重要的, 关键的, 批判的	11
cross-sectional area 横截面面积	R17
cross-sectional data 横截面数据	R17
crucial [ˈkruːʃl] adj. 关键的, 至关重要的, 关键性的	R9
crude [kruːd] adj. 粗略的, 简略的, 大概的	R4
crystal display panel 液晶显示面板	15
crystalline [ˈkrɪstlaɪn] adj. 水晶(般)的	21
cumulative [ˈkjuːmjələtɪv] adj. 聚积的, 积累的, 渐增的	6
custom shape 自定义形状	R18
cutting force 切削力	23
cyber-physical system 信息物理系统	2
cylindrical [səˈlɪndrɪkl] adj. 圆柱体的	6
cylindrical coordinate robot 圆柱坐标型机器人	6

D

data compilation 数据汇编	R16
data mining 数据挖掘	R21
data traffic 数据流	R20
database [ˈdeɪtəbeɪs] n. (储存在计算机中的)数据库	R9
DC generator 直流发电机	23
DC relaxation circuit 直流弛张电路	23
debug [ˈdiːbʌɡ] v. 排错, 调试	10

deceleration [ˌdiːseləˈreɪʃn] n. 减速,降速   6
decentralize [ˌdiːˈsentrəlaɪz] v. 分散,分权,使(业务)分散,疏散(人口)   2
dedicated [ˈdedɪkeɪtɪd] adj. 专用的,专门用途的,献身的,专心的   11
dedicated bus 专用总线   R12
dedication [ˌdedɪˈkeɪʃn] n. 投入,奉献精神   R14
deep learning 深度学习   3
deep penetration welding 深熔焊   22
deep space probe mission 深空探测任务   R22
defect [ˈdiːfekt] n. 缺点,缺陷,毛病   19
defect detection algorithm 缺陷检测算法   19
defective [dɪˈfektɪv] adj. 有问题的,有缺陷的   R15
definition [ˌdefɪˈnɪʃn] n. 清晰,清晰度,定义,释义,榜样,典范   11
degree of freedom 自由度   6
delegate [ˈdelɪgeɪt] v. 授(权),把(职责、责任等)委托(给)   2
delivery [dɪˈlɪvəri] n. 递送,投递,分娩,生产,演讲风格   26
delivery projection 交付预测   R14
delta robot 三角机器人   6
demand [dɪˈmɑːnd] n. 需求,需求量 v. 强烈要求,需要,需求   26
drop-on-demand approach 随需应变的方法   R17
densely [ˈdensli] adv. 稠密地,密集地   R17
dependency [dɪˈpendənsi] n. 依靠,依赖   R2
deploy [dɪˈplɔɪ] v. 部署,有效利用   20
deployment [dɪˈplɔɪmənt] n. 部署,调集   R13
deposit [dɪˈpɒzɪt] v. 放置,使沉积,使沉淀   R17
deposition [ˌdepəˈzɪʃn] n. 沉淀,罢免,废黜   24
depress [dɪˈpres] v. 按,压,推下(尤指机器部件)   15
derivative [dɪˈrɪvətɪv] adj. 衍生的   3
derivative design 衍生式设计   3
derived [dɪˈraɪvd] adj. 导出的,衍生的,派生的   16
derived signal 衍生信号   16
description [dɪˈskrɪpʃn] n. 说明,描述,形容   25
design [dɪˈzaɪn] n. 设计,布局,安排   26
designate [ˈdezɪgneɪt] v. 指定,认定,委任,标明,(已当选)尚未就职   30
destination [ˌdestɪˈneɪʃn] n. 目的地,终点   14
detect [dɪˈtekt] v. 查明,检测,发现,察觉   3

detection[dɪˈtekʃn] n. 检测，察觉，发现	3
detergent [dɪˈtɜːdʒnt] n. 洗涤剂，去污剂	8
deterministic [dɪˌtɜːmɪˈnɪstɪk] adj. 基于决定论的，不可抗拒的，不可逆转的	R12
deterministic time delay 确定性时间延迟	R20
device [dɪˈvaɪs] n. 装置，仪器，器具，设备	2
device shadow 设备影子	12
devolve [dɪˈvɒlv] v. 转移，移交，(使)(权力、职责等)下放	R9
diagnostic reasoner(s) 诊断推理器	R2
diameter [daɪˈæmɪtə(r)] n. 直径，放大倍数	25
dielectric [ˌdaiiˈlektrik] adj. 非传导性的，诱电性的	23
dielectric fluid 绝缘液体	23
differential [ˌdɪfəˈrenʃl] adj. 微分的，差别的，特异的	18
differential equation 微分方程	18
diffraction [dɪˈfrækʃn] n. 衍射	R18
digital [ˈdɪdʒɪtl] adj. 数字的，数码的，数字信息系统的	1
digital companion 数字伙伴	12
digital mapping 数字映射	12
digital mirroring 数字镜像	12
digital radiographic detection system 数字射线检测系统	19
digital radioscopy 数字放射镜	19
digital twin 数字孪生	2
digital-to-analog converter (DAC) 数模转换	R12
digitize [ˈdɪdʒɪtaɪz] v. 使数字化	2
dimension[dɪˈmenʃn] n. 尺寸	R6
dimensional [daɪˈmenʃən(ə)l] adj. 维度的	R22
dip-pen nanolithography 蘸笔纳米光刻	R18
direct digital manufacturing (DDM) 直接数字化制造	24
disassembly [ˌdɪsəˈsembli] n. 拆卸，分解	6
discharge [dɪsˈtʃɑːdʒ] v. 放电	23
discipline [ˈdɪsəplɪn] n. (尤指大学的)科目，学科，纪律，自制力	R21
disconnect [ˌdɪskəˈnekt] v. 切断，使分离	8
dismantle [dɪsˈmæntl] v. 拆开，拆卸	R3
dispatch [dɪˈspætʃ] v. 调度，发送，迅速处理	17
dispense [dɪˈspens] v. 分配，分发，施与	R17
disperse [dɪˈspɜːs] v. 分散，散布，疏散，驱散	3

displace [dɪs'pleɪs] v. 取代,替代,置换   2

displaced [dɪs'pleɪst] adj. 位移的,被取代的,无家可归的   19

displaced signal 位移信号   19

displacement [dɪs'pleɪsmənt] n. 移位,取代   6

disposal [dɪ'spəʊzl] n. 处理,清除   27

dispose [dɪ'spəʊz] v. 处理,安排,丢弃   27

dissolve [dɪ'zɒlv] v. (使)溶解,解散,解除,消失   21

distinguish [dɪ'stɪŋgwɪʃ] v. 区分,辨别,分清   R4

distortion [dɪ'stɔːʃn] n. 失真,扭曲,歪曲   R10

distortion rate 失真率   22

distributed [dɪ'strɪbjuːtɪd] adj. 分布式的   1

distributed integration 分布式集成   1

distributed network 分布式网络   R13

diverse [daɪ'vɜːs] adj. 不同的,各式各样的   R23

dock [dɒk] v. 对接,(船舶)进坞   9

documentation [ˌdɒkjumen'teɪʃn] n. 文件,凭证   R7

domain [də'meɪn] n. 域,定义域   R2

dominant ['dɒmɪnənt] adj. 显性的,占优势的,支配的,统治的   22

double closed-loop control 双闭环控制算法   30

downlink ['daʊnlɪŋk] n. 下行链路   11

downstream [ˌdaʊn'striːm] adj. 顺流的,在(或向)下游的   R23

downtime management 停机管理   17

downtime ['daʊntaɪm] n. (尤指计算机的)停机时间,停止运行时间   2

draft [drɑːft] n. 草稿,草案,草图   R10

drain bias voltage 漏极偏置电压   21

drain diffusion 泄漏扩散   21

draw upon 利用   R1

drive module 驱动模块   R7

driver's license 驾照   R9

dual cabinet controller 双机柜控制器   R7

dual-frequency receiver 双频接收器   R11

dual polarity 双极性   14

duplicate ['djuːplɪkeɪt, 'djuːplɪkət] adj. 重复的,完全一样的,复制的,副本的   R16

durability [ˌdjʊərə'bɪlɪti] n. 耐久性   6

durable ['djʊərəbl] adj. 耐用的,持久的   R5

dynamic [daɪˈnæmɪk] *adj*.动态的 *n*.动力,动力学　　R9
dynamic alliance 动态联盟　　26
dynamically [daɪˈnæmɪkli] *adv*.动态地　　13

E

ecological [ˌiːkəˈlɒdʒɪkl] *adj*.生态的,生态学的　　27
ecological law 生态规律　　27
economics [ˌiːkəˈnɒmɪks] *n*.经济学,经济情况　　R1
effector [ɪˈfektə(r)] *n*.效应器　　6
effector organ 操纵机构　　6
e-government 电子政务　　R9
e-health 电子健康　　R9
electric motor 电动机　　23
electric shock 电击　　10
electrical [ɪˈlektrɪk(ə)l] *adj*.电的,与电有关的　　23
electrical cabinet 电器柜　　10
electrical discharge machining (EDM) 电火花加工　　23
electrical energy 电能　　23
electrocardiograph [ɪˌlektrəʊˈkɑːdiəɡrɑːf] *n*.心电图　　R10
electrode [ɪˈlektrəʊd] *n*.电极　　23
electrolyze [ɪˈlektrəʊlaɪz] *v*.电解　　21
electromagnetic [ɪˌlektrəʊmæɡˈnetɪk] *adj*.电磁的　　4
electromagnetic field 电磁场　　4
electromagnetic interference 电磁干扰　　R20
electromagnetic interrogation pulse 电磁询问脉冲　　4
electron [ɪˈlektrɒn] *n*.电子　　24
electron-beam lithography 电子束光刻　　R18
electron-beam direct-write lithography 电子束直写光刻　　R18
electronics [ɪˌlekˈtrɒnɪks] *n*.电子技术,电子学　　2
electron-sensitive film 电子敏感膜　　R18
eliminate [ɪˈlɪmɪneɪt] *v*.消除,排除,清除　　R16
embed [ɪmˈbed] *v*.嵌入　　4
emergency [ɪˈmɜːdʒənsi] *n*.突发事件,紧急情况　　R7
emergency by-pass plug 紧急旁通插头　　15
emergency stop button 紧急停止按钮　　R7
emission [ɪˈmɪʃn] *n*.排放物,散发物　　27

empower [ɪmˈpaʊə(r)] v. 授权,给予(某人)……权力 　　　　　　　　R9
emulate [ˈemjuleɪt] v. 仿真,模仿 　　　　　　　　R2
encapsulate [ɪnˈkæpsjuleɪt] v. 压缩,简述,概括 　　　　　　　　R3
encompass [ɪnˈkʌmpəs] v. 包含,涉及(大量事物),包围,围绕 　　　　　　　　11
encrypted [ɪnˈkrɪptɪd] adj. 加密的 　　　　　　　　R11
encrypted P(Y) code 加密码 　　　　　　　　R11
encryption [ɪnˈkrɪpʃn] n. 加密,加密技术 　　　　　　　　11
endurance [ɪnˈdjʊərəns] n. 耐久力,忍耐力 　　　　　　　　R5
energy consumption 能源消耗 　　　　　　　　3
enhance [ɪnˈhɑːns] v. 增强,提高,改善 　　　　　　　　R12
entirely [ɪnˈtaɪəli] adv. 完全地,全部地,完整地 　　　　　　　　R9
entity [ˈentəti] n. 实体 　　　　　　　　1
environment [ɪnˈvaɪrənmənt] n. 周围状况,条件,工作平台,自然环境 　　　　　　　　26
equator [ɪˈkweɪtə(r)] n. 赤道 　　　　　　　　R11
equivalent [ɪˈkwɪvələnt] adj. 相等的,相同的 n. 相等的东西,等量 　　　　　　　　9
eradication [ɪˌrædɪˈkeɪʃn] n. 消除,根除,消灭 　　　　　　　　17
erasable optical media 可擦除光媒体 　　　　　　　　R10
erase [ɪˈreɪz] v. 擦掉,抹掉,删除 　　　　　　　　15
erosion [ɪˈrəʊʒən] n. 腐蚀(作用),磨损 　　　　　　　　23
estimate [ˈestɪmət, ˈestɪmeɪt] v. 估计 　　　　　　　　10
estimation [ˌestɪˈmeɪʃn] n. 估计,判断,评价 　　　　　　　　R10
e-stop 紧急停工 　　　　　　　　16
etch [etʃ] v. 蚀刻,凿出 　　　　　　　　R18
etching [ˈetʃɪŋ] n. 蚀刻 　　　　　　　　21
ethernet [ˈiːθənet] n. 以太网 　　　　　　　　R12
evacuate [ɪˈvækjueɪt] v. 疏散,撤离 　　　　　　　　R19
evaluate [ɪˈvæljueɪt] v. 评价,评估,估值 　　　　　　　　19
evaporate [ɪˈvæpəreɪt] vt. 使……蒸发,使……脱水,使……消失 　　　　　　　　22
evident [ˈevɪdənt] adj. 显然的,清楚的,显而易见的 　　　　　　　　30
exceed [ɪkˈsiːd] v. 超过 　　　　　　　　10
excessive [ɪkˈsesɪv] adj. 过度的,过多的 　　　　　　　　7
exclusive [ɪkˈskluːsɪv] adj. 专用的,专有的,独有的,独占的 　　　　　　　　R4
exclusively [ɪkˈskluːsɪvli] adv. 唯一地,排他地,独占地 　　　　　　　　R10
executing capacity plan 执行产能计划 　　　　　　　　R14
executing material plan 执行物料计划 　　　　　　　　R14

exothermic [ˌeksə(ʊ)ˈθɜːmɪk] adj. 发热的，放热的	22
exothermic reaction 放热反应	22
expel [ɪkˈspel] vt. 驱逐，开除	22
experimentally [ɪkˌsperəˈmentlɪ] adv. 用实验方法地，实验式地	R10
explicitly [ɪkˈsplɪsɪtlɪ] adv. 清楚明确地，详述地	R21
exploration [ˌekspləˈreɪʃən] n. 探索，研究	5
exploratory [ɪkˈsplɒrət(ə)rɪ] adj. 探究的，勘探的，考察的	R21
exponential [ˌekspəˈnenʃl] n. 指数函数 adj. 指数的，迅猛的，呈几何级数的	2
exposure [ɪkˈspəʊzə(r)] n. 暴露	4
external [ɪkˈstɜːnl] adj. 外部的，外面的	10
external impulse 外界冲击强度	5
extinguishing [ɪkˈstɪŋgwɪʃɪŋ] n. 熄灭 v. 熄灭，(使)消亡，破灭	R7
extract [ɪkˈstrækt] v. 提取，提炼	5
extreme ultraviolet lithography 极紫外光刻	R18
extrusion [ɪkˈstruːʒn] n. 挤压	24

F

fabrication [ˌfæbrɪˈkeɪʃn] n. 制作，制造，建造，装配	21
fabrication process 制造工艺	21
facility [fəˈsɪlətɪ] n. 设施，设备	2
facsimile [fækˈsɪməlɪ] n. 传真，复写	R22
Fahrenheit [ˈfærənhaɪt] n. 华氏温度(计)(的)，华氏(温标)	23
fatigue [fəˈtiːg] n. 疲乏，厌倦，(金属部件的)疲劳	19
fault [fɔːlt] n. 故障，过失，缺点，缺陷	3
fault diagnosis 故障诊断	1
fault tolerance 容错性	30
feasibility feedback 可行性反馈	R13
feed hold 持续给进	16
feed override 速度倍率设定	16
feed piston 馈料活塞	R17
feed rate 进给速率	R19
feed speed 进给速度	25
feedback [ˈfiːdbæk] n. 反馈，反馈的意见(或信息)	3
fertilizer [ˈfɜːtəlaɪzə(r)] n. 化肥，肥料	R8
fiber delivery 纤维输送	22
file server 文件服务器	R13

filing [ˈfaɪlɪŋ] n. 锉削	18
filling hole 充油孔	7
filling plug 充油塞	7
film-covered packaging box 覆膜包装盒	R8
filter [ˈfɪltə(r)] n. 过滤器,滤声器,滤波器	8
filter holder 过滤器座	8
filtering operation 滤波操作	R10
finance [ˈfaɪnæns] n. 财务,资金,财政,金融	R14
finish [ˈfɪnɪʃ] v. 完成,做好 n. 修整,精修,精整,精制,修正	23
finished part 成品零件	3
finite number 有限数	R22
fixture [ˈfɪkstʃə(r)] n. 固定装置,夹具	15
flange [flændʒ] n. 凸缘	R6
flaw [flɔː] n. 错误,缺点	R4
flexibility [ˌfleksəˈbɪləti] n. 灵活性,弹性,柔性	16
flexible [ˈfleksəbl] adj. 灵活的,易变通的,适应性强的	1
flexible factory 柔性工厂	11
flexible manufacturing system 柔性制造系统	R20
flexibly [ˈfleksəbli] adv. 灵活地,柔软地,有弹性地	R8
FlexPendant 示教器	R7
flowchart [ˈfləʊtʃɑːt] n. 流程图	R15
fluctuate [ˈflʌktʃueɪt] v. 波动,起伏不定,使波动	17
fluid [ˈfluː(ː)ɪd] n. 流体,液体	23
fluorine [ˈflɔːriːn] n. 氟	21
fluorine gas 氟气	21
focus [ˈfəʊkəs] v. 集中,关注,聚焦,调焦	26
forefront [ˈfɔːfrʌnt] n. (运动、活动的)前沿,(思考、关注的)重心	27
foreman [ˈfɔːmən] n. 领班,陪审团主席	16
formal logic 形式逻辑	R1
formidable [fəˈmɪdəb(ə)l] adj. (形状、数量)巨大的,可怕的,令人敬畏的	26
fossil fuel 化石燃料	27
fraction [ˈfrækʃn] n. 小部分,微量,分数,小数	R19
framework [ˈfreɪmwɜːk] n. 框架,(建筑物或物体的)构架	R10
free tolerance 自由公差	25
freight board 运货板	R8

frequency [ˈfriːkwənsi] n. 频繁，频率	4
friction [ˈfrɪkʃ(ə)n] n. 摩擦，摩擦力，不和，分歧	R19
functionality [ˌfʌŋkʃəˈnæləti] n. 功能性	2
fundamental [ˌfʌndəˈmentl] adj. 根本的，基本的	R16
fuse [fjuːz] v. 熔化，融合	24
fusion [ˈfjuːʒn] n. 融合，结合	12
fuzzy [ˈfʌzi] adj. (图片、声音等)不清楚的，模糊的	19
fuzzy algorithm 模糊算法	19
fuzzy control technology 模糊控制技术	1
fuzzy recognition theory 模糊识别理论	19

G

gantry [ˈgæntri] n. 桁架，桶架	6
gantry robot 桁架机器人	6
gap [gæp] n. 间隙，间隔，间断，空白，空隙，距离，范围	23
gear [gɪə(r)] n. 齿轮，传动装置	R6
gearbox [ˈgɪəbɒks] n. 齿轮箱	7
generale [ˈdʒenəreɪt] v. 产生，引起	18
generator [ˈdʒenəreɪtə] n. 发电机，产生者	23
geographically [ˌdʒiːəˈgræfɪkli] adv. 在地理上，地理学上	R21
geometric [ˌdʒiːəˈmetrɪk] adj. 几何图形的，几何的	R23
geometric tolerance 形位公差	25
geometry [dʒiˈɒmətri] n. 几何图形，几何结构	24
global positioning system (GPS) 全球定位系统	R11
glue [gluː] n. 黏合剂，胶水	R16
grabber [ˈgræbə(r)] n. 采集器，掠夺者	19
graphic [ˈgræfɪk] n. 图表，图形，图画	R10
graphic display 图形显示	R15
graphical [ˈgræfɪkl] adj. 图形的，用图(或图表等)表示的	R12
graphical user interface 图形化用户接口	R13
grasp [grɑːsp] v. 抓牢，握紧，理解，领会，抓住(机会)	26
grease [griːs] n. 油脂，润滑油	7
grease nipple 润滑脂头	7
grease pump 润滑脂泵	7
green manufacture 绿色制造	27
greenhouse [ˈgriːnhaʊs] n. 温室，暖房 adj. 温室效应的	27

greenhouse gas 温室气体（二氧化碳、甲烷等） 27
gripper [ˈɡrɪpə(r)] n. 夹具，钳子 R8
guarantee [ˌɡærənˈtiː] n. 保证，担保 R23
guidance [ˈɡaɪdns] n.（火箭等的）制导，指导，指引 R22
gyroscope [ˈdʒaɪrəskəʊp] n. 陀螺仪 21

H

habitat [ˈhæbɪtæt] n. 栖息地 27
hand-held 手持式的 15
hardness [ˈhɑːdnɪs] n. 硬度，刚度，强度，(坚)硬性，刚度指数 23
hardware abstraction system 硬件抽象系统 R12
hazardous [ˈhæzədəs] adj. 危险的，有害的，碰运气的 21
helmet [ˈhelmɪt] n. 头盔，防护帽 10
heterogeneous [ˌhetərəˈdʒiːniəs] adj. 异质的，不均一的，参差的 26
heterogeneous distribution environment 异构分布环境 26
hexagon [ˈheksəɡən] n. 六边形，六角形 R5
hexagon socket head cap screw 内六角圆柱头螺钉 R5
high-resolution 高清晰度的，高分辨率的 R22
high-speed machining 高速加工 R19
histogram [ˈhɪstəɡræm] n. 直方图，柱状图 R15
holding brake 保持制动器 R7
hollow [ˈhɒləʊ] adj. 空的，凹陷的，虚伪的 7
hollow wrist 空心腕关节 7
holographic [ˌhɒləˈɡræfɪk] adj. 全息图的 13
holographic technology 全息技术 13
holonic [həʊˈlɒnɪk] adj. 整子的，子整体的，合子 1
holonic system 整子系统 1
horizontal [ˌhɒrɪˈzɒnt(ə)l] adj. 水平的，统一的 7
horizontal arm 水平臂 7
horizontally [ˌhɒrɪˈzɒntəli] adv. 水平地，横地 6
human-robot collaboration 人机协作 30
hybrid [ˈhaɪbrɪd] adj. 混合的 n. 合成物 9
hybrid manufacturing 混合制造 R23
hybrid robot 混合机器人 9
hypermedia [ˌhaɪpəˈmiːdiə] n. 超媒体 R10
hypothetical [ˌhaɪpəˈθetɪkl] adj. 假设的，假定的 R2

I

identify [aɪˈdentɪfaɪ] v. 确认，认出，鉴定	2
identifying inventory number 识别库存编号	4
identity [aɪˈdentətɪ] n. 身份，本体，个性，特性	R9
idle [ˈaɪdl] adj. 空闲的，闲置的	20
if-then rule 如果-那么规则	R2
ignition [ɪgˈnɪʃ(ə)n] n. 点火，点燃，着火	22
ignition temperature 燃点	22
illustration [ˌɪləˈstreɪʃ(ə)n] n. 插图，说明，实例	8
image [ˈɪmɪdʒ] n. 影像，形象，印象，声誉	3
image grabber 图像采集器	19
image preprocessing 图像预处理	19
imaginary [ɪˈmædʒɪnərɪ] adj. 想象中的，幻想的，虚构的	R12
impact [ˈɪmpækt] n. 巨大影响，强大作用，撞击，冲击力	27
imperative [ɪmˈperətɪv] adj. 必要的，命令的，强制的	19
implant [ɪmˈplɑːnt] n. 植入物	24
implementation [ˌɪmplɪmenˈteɪʃən] n. 履行，实施	1
impulse [ˈɪmpʌls] n. 脉冲，冲动	5
in accordance with 符合，依照，和……一致	R5
in the infrared range of 在红外线范围内	22
inadvertently [ˌɪnədˈvɜːtəntlɪ] adv. 无意地，不经意地	10
incompatible [ˌɪnkəmˈpætəbl] adj. 不兼容的，不相容的	R16
inconsistent [ˌɪnkənˈsɪstənt] adj. 不一致的，不协调的，前后矛盾的	19
incremental [ˌɪŋkrəˈment(ə)l] adj. 增加的，递增的，逐渐的	25
incremental value instruction 增量值指令	25
incrementally [ˌɪŋkrɪˈmentəlɪ] adv. 递增地，逐渐地，增加地	R17
incumbent [ɪnˈkʌmbənt] adj. 在职的，现任的	2
indication [ˌɪndɪˈkeɪʃn] n. 指示，象征，迹象	16
indicator [ˈɪndɪkeɪtə(r)] n. 标志，迹象，指标	17
induction [ɪnˈdʌkʃn] n. 归纳，归纳法，引导	R23
industrial big data 工业大数据	R23
industrial ecology 工业生态学	27
inert gas 惰性气体	22
inference [ˈɪnfərəns] n. 推断，推理，推论	R2
information flow 信息流	29

information modeling 信息建模	12
information platform 信息平台	12
information repository 信息存储库	R16
infrared [ˌɪnfrəˈred] n. 红外线 adj. 红外线的	22
infrastructure [ˈɪnfrəstrʌktʃə(r)] n. 基础设施	R13
inherently [ɪnˈherəntli] adv. 内在地, 固有地	19
initialization [ɪˌnɪʃəlaɪˈzeɪʃn] n. 初始化	20
injured [ˈɪndʒəd] adj. 受伤的, 有伤的, 委屈的	R7
inkjet [ˈɪŋkˌdʒet] n. 喷墨	R17
innovation [ˌɪnəˈveɪʃn] n. (新事物、思想或方法的)创造, 创新, 改革	2
input symbol 输入符号	5
insert [ɪnˈsɜːt] v. 插入, 嵌入, (在文章中)添加	8
installation [ˌɪnstəˈleɪʃn] n. 安装, 设置, 装置, 设备	R7
instantaneous [ˌɪnstənˈteɪniəs] adj. 立即的, 立刻的, 瞬间的	R4
instantaneous response 瞬间响应	R20
instrument [ˈɪnstrəmənt] n. 器械, 仪器, 器具	R12
insulation [ˌɪnsjuˈleɪʃ(ə)n] n. 隔热, 绝缘	10
insulation protection cover 绝缘保护罩	10
insulator [ˈɪnsjʊleɪtə] n. 绝缘体	21
intact [ɪnˈtækt] adj. 原封不动的, 完整的	21
integrate [ˈɪntɪɡreɪt] v. 集成, 整合, (使)合并	R14
integrated [ˈɪntɪɡreɪtɪd] adj. 综合的, 完整统一的, 各部分密切协调的	1
integrated circuit 集成电路	19
integrated intelligent system 一体化智能系统	1
intelligent [ɪnˈtelɪdʒənt] adj. 聪明的, 有智力的, 智能的	5
intelligent embedding 智能嵌入	R3
intelligent manufacturing 智能制造	2
intensifier [ɪnˈtensɪfaɪə(r)] n. 增强器	19
interaction [ˌɪntərˈækʃən] n. 相互影响, 相互作用	4
interactive multimedia 交互式多媒体	R10
interconnection [ˌɪntəkəˈnekʃn] n. 互联互通, 紧密联系, 关联	2
inter-connectivity 互联性	2
interface [ˈɪntəfeɪs] n. 界面, 接口	20
interference [ˌɪntəˈfɪərəns] n. 干涉, 干预, 介入	2
interlock [ˌɪntəˈlɒk] v. (使)连锁	R5

intermediate [ˌɪntəˈmiːdiət] adj. 中间的,居中的,中级的	17
internal sleeve 内部套筒	7
internet of people 人联网	2
internet of things 物联网	2
interpersonal [ˌɪntəˈpɜːsənl] adj. 人际关系的,人际的	27
interpolation [ɪnˌtɜːpəˈleɪʃ(ə)n] n. 插入,篡改,填写,插值	25
interpreter [ɪnˈtɜːprɪtə] n. 读卡机,解释程序,口译译员,演绎者,表演者	R12
interrogating radio wave 询问无线电波	4
interrogation [ɪnˌterəˈgeɪʃn] n. 询问	4
intervention [ˌɪntəˈvenʃn] n. 干涉,干预	3
intuitive [ɪnˈtjuːɪtɪv] adj. 直观的,直觉的,凭直觉得到的,有直觉力的,易懂的	9
inventory [ˈɪnvəntri] n. 存货,库存	4
inventory file 库存文件	R14
inventory goods 库存	4
inventory optimization 库存优化	R16
inventory record 库存记录	29
investment [ɪnˈvestmənt] n. 投资	26
invoke [ɪnˈvəʊk] v. 引起,提及,调用,激活	28
ionize [ˈaɪənaɪz] v. (使)电离,离子化	21
ionized metal vapor 电离金属蒸气	22
ionosphere [aɪˈɒnəsfɪə(r)] n. 电离层	R11
islands of automation 自动化孤岛	29
isolated [ˈaɪsəleɪtɪd] adj. 单独的,孤寂的,遥远的,偏僻的	R23
isolator [ˈaɪsleɪtə(r)] n. 隔离器,隔音装置,绝缘体	19
issue [ˈɪʃuː] v. 下发,宣布,公布	20
iteration [ˌɪtəˈreɪʃn] n. 迭代	3

J

joint [dʒɔɪnt] n. 关节	15
just-in-case inventory 即时库存	17

K

keyhole welding 穿透型焊接法	22
keypad [ˈkiːpæd] n. 按键	15
kinetic [kɪˈnetɪk] adj. 运动的,活跃的	22
kinetic energy 动能	22

L

label ['leɪbl] n.标签 v.贴标签于,用标签标明	4
labor productivity 劳动生产率	R8
large-scale [ˌlɑːdʒ'skeɪl] adj.大规模的,大范围的	R23
laser ['leɪzə(r)] n.激光,激光器	14
laser guided vehicle 激光导引车	14
laser scanner 激光扫描器	9
laser target navigation 激光目标导航	14
latency ['leɪtənsi] n.时延	11
lathe [leɪð] n.车床,机床	20
latitude ['lætɪtjuːd] n.纬度,纬度地区	R11
launch [lɔːntʃ] n.(产品的)上市	3
layout ['leɪaʊt] n.布局,布置	3
lead [liːd] v.引领,带路 n.领先地位	19
lead plate area 导板区	19
lead time 订货交付时间	11
lean [liːn] adj.精简的,效率高的 v.倚靠,靠在	26
lean manufacturing 精益生产	26
legitimacy [lɪ'dʒɪtɪməsi] n.合法性,正统	R16
lens ['lenz] n.透镜	22
leverage ['liːvərɪdʒ] v.充分利用(资源、观点等) n.影响力,手段,杠杆作用	R21
life cycle 生命周期	12
light coupling mask nanolithography 光耦合掩模纳米光刻	R18
linear ['lɪniə(r)] adj.线性的,直线的	25
linear cutting instruction 直线切削指令	25
linear economy 线性经济	27
linguistics [lɪŋ'gwɪstɪks] n.语言学	R1
lint [lɪnt] n.棉绒,毛絮,线头	7
lint-free cloth 无绒布	7
liquid binder 液态黏合剂	R17
liquid immersion 液浸	R18
lithography [lɪ'θɒɡrəfi] n.刻蚀术	R18
load [ləʊd] n.负载,负荷,装载量 v.承载,装入	3
load sensor 负载传感器	30
local oxidation nanolithography 局部氧化纳米光刻	R18

locate [ləʊˈkeɪt] v. 确定……的位置, 设立, 建立	8
login prompt 登录提示	20
login setting 登录设置	20
logistics [ləˈdʒɪstɪks] n. 物流, 后勤学, 运筹学, 统筹安排	2
longitude [ˈlɒŋɡɪtjuːd] n. 经度	R11
lot [lɒt] n. 批次, 小块土地, 电影摄制场	17
low latency 低时延	11
lubricate [ˈluːbrɪkeɪt] v. 润滑, 给……加润滑油, 促进	7
lubrication point 润滑点	7

M

machine learning 机器学习	R21
machine tool 机床	3
machine vision 机器视觉	9
machine wear 机床磨损	R19
macroscopic [ˌmækrəˈskɒpɪk] adj. 宏观的	21
magnet [ˈmæɡnət] n. 磁铁, 磁石, 有吸引力的人(或地方、事物), 磁体	14
magnetic [mæɡˈnetɪk] adj. 磁的, 磁性的	14
magnetic tape 磁带	14
magnitude [ˈmæɡnɪtjuːd] n. 巨大, 重大	R10
main power switch 主电源开关	R7
mainstream [ˈmeɪnstriːm] adj. 主流的	R10
maintain [menˈteɪn] v. 维护, 保养, 支持, 保持	23
maintenance [ˈmeɪntənəns] n. 维护, 保养	2
malfunction [ˌmælˈfʌŋkʃn] n. 故障, 失灵	R20
maneuver [məˈnuːvə] v. (熟练地)移动, 调动, 转动, 操纵	6
manipulate [məˈnɪpjuleɪt] v. (熟练地)操作, 使用	R1
manipulator [məˈnɪpjuleɪtə(r)] n. 机械手, 调制器, 操作者	7
manipulator arm 机械手臂	6
manipulator system 机械手系统	R7
manual [ˈmænjuəl] n. 使用手册, 风琴键盘 adj. 手动的, 手工的, 手控的	7
manual programming 手工编程	25
manually [ˈmænjuəli] adv. 手动地, 用手	R7
manufacturability [ˌmænjʊfæktʃərəˈbɪlɪtɪ] n. 可制造性, 工艺性, 可生产性	12
manufacturing [ˌmænjuˈfæktʃərɪŋ] n. 制造	2
manufacturing resource 制造资源	18

mapping ['mæpɪŋ] n. (数学、语言学)映射,映现   12
marketing ['mɑːkɪtɪŋ] n. 促销,营销   R14
marketplace ['mɑːkɪtpleɪs] n. 市场,集市   13
mass production 大批量生产   30
massive ['mæsɪv] adj. 大量的,大规模的   26
master scheduling 主调度   R14
master-slave structure 主从结构   R13
material level initialization 料位初始化   20
mathematical optimization 数学优化   R1
matrix ['meɪtrɪks] n. 矩阵,模型   R22
mechanical [məˈkænɪkl] adj. 机械的,机械驱动的,呆头呆脑的,机械学的   9
mechanical skeleton 机械框架,机械骨架   5
mechanism ['mekənɪzəm] n. 机构,结构,构造,机械装置   R2
mechatronics [ˌmekəˈtrɒnɪks] n. 机械电子学   9
message routing 消息路由   R1
metal removal 金属切削   23
metallurgical [ˌmetəˈlɜːdʒɪkəl] adj. 冶金的,冶金学的   23
metallurgical change 金相组织变化   23
methodology [ˌmeθəˈdɒlədʒi] n. 方法,原则,制备工艺   28
metric ['metrɪk] n. 指标,衡量标准,度规   16
microelectronics [ˌmaɪkrəʊɪˌlekˈtrɒnɪks] n. 微电子技术,微电子学   9
micromachining [ˌmaɪkrəʊməˈʃiːnɪŋ] n. 显微机械加工,微细加工   21
microscopic [ˌmaɪkrəˈskɒpɪk] adj. 微观的   21
middleware ['mɪdlweə(r)] n. 中间件,中介软件(允许不同程序协同工作)   R1
milling ['mɪlɪŋ] n. 铣削   18
milling depth 铣切深度   25
milling system 铣削系统   R19
mimic ['mɪmɪk] v. 模仿(人的言行举止),模拟   R10
mine [maɪn] v. 采(煤等矿物),挖掘 pron. 我的 n. 矿,矿井,宝库,源泉   27
minimize ['mɪnɪmaɪz] v. 使减小到最低限度,使最小化   27
minor ['maɪnə(r)] adj. 较小的,次要的   26
mirroring ['mɪrərɪŋ] n. 反射,镜面反射   12
miscellaneous [ˌmɪsəˈleɪniəs] adj. 各种各样的,混杂的   R18
mission-critical 关键的,至关重要的   11
mixed manufacturing 混合制造   R23

词条	页码
model ['mɒdl] n. 样式,设计,模型	3
model based definition 基于模型的定义	R19
model interaction 模型交互	12
model-based enterprise 基于模型的企业	R23
moderate-throughput 中等吞吐量	R12
moderate ['mɒdəreɪt] adj. 适度的,中等的,温和的	R12
modernization [ˌmɒdənaɪ'zeɪʃn] n. 现代化	R8
modernization process 现代化进程	R8
modification [ˌmɒdɪfɪ'keɪʃn] n. 修改,更改	R20
modify ['mɒdɪfaɪ] v. 修改,修饰,限定,使温和	18
modular ['mɒdjələ(r)] adj. 模块化的,组合式的	R12
modularity [ˌmɒdjʊ'lærɪtɪ] n. 模块化,模块性	12
modulated ['mɔdjuleitid] adj. 已调的,被调的	25
modulated aluminum 调制铝	25
module ['mɒdjuːl] n. 模块,功能块,程序块,组件,配件,舱,单元	9
mold [məʊld] n. 模具,铸模	R19
molten metal 熔融金属	22
monitor ['mɒnɪtə(r)] v. 监视,检查,跟踪调查	2
monitoring system 监控系统	17
monotonous [mə'nɒtənəs] adj. 单调乏味的,毫无变化的	R8
morale [mə'rɑːl] n. 士气,精神面貌	27
mortal ['mɔːtl] adj. 不能永生的,终将死亡的	R2
motion sensor 运动传感器	21
motivate ['məʊtɪveɪt] v. 激励,激发,成为……的动机	27
motor ['məʊtə(r)] n. 发动机,引擎	R6
mount [maʊnt] v. 安装,安置,裱	R6
multi-agent system 多智能体系统	1
multidimensional [ˌmʌltidaɪ'menʃənl] adj. 多维的	18
multimedia [ˌmʌlti'miːdiə] n. 多媒体	13
multiphoton lithography 多光子光刻	R18
multiple ['mʌltɪpl] adj. 数量多的,多种多样的	1
multiple-electron beam 多电子束	R18
multi-sensor fusion 多传感器融合	13
multivariate [ˌmʌltɪ'veərɪt] adj. 多元的,多变量的	12
mutual ['mjuːtʃuəl] adj. 相互的,彼此的,共同的,共有的	R23

myriad [ˈmɪriəd] n. 大量,无数	R16

N

nanopattern graphene 纳米图形石墨烯	R18
nanotechnology [ˌnænəʊtekˈnɒlədʒi] n. 纳米技术	R9
native code 本地代码	R12
navigation [ˌnævɪˈɡeɪʃn] n. 导航,领航,航行	14
navigation system gyroscope 导航系统陀螺仪	21
negative direction 负向	15
negligible [ˈneɡlɪdʒəbl] adj. 可以忽略不计的,微不足道的,不重要的	R17
negotiate [nɪˈɡəʊfieɪt] v. 商定,达成协议,谈判,洽谈	26
neural [ˈnjʊərəl] adj. 神经的,神经系统的	R21
neural network 神经网络	3
neuron [ˈnjʊərɒn] n. 神经元	R10
niche [niːʃ] n. 市场定位,生态位	R10
nipple [ˈnɪpl] n. 乳头,奶嘴	7
nitrogen [ˈnaɪtrədʒ(ə)n] n. 氮气	22
node [nəʊd] n. 节点,结点,茎节	R20
nonconducting fluid 非导电流体	23
nondestructive [ˌnʌndɪˈstrʌktɪv] adj. 无损的	3
nondestructive testing 无损检测	3
nonetheless [ˌnʌnðəˈles] adv. 尽管如此,不过	R14
nonlinear process 非线性过程	R10
non-reactive gas 非反应性气体	22
notification [ˌnəʊtɪfɪˈkeɪʃn] n. 通知,通告,布告	16
nozzle [ˈnɒzl] n. 喷嘴	R17
nuanced [ˈnjuːɒnst] adj. 微妙的,具有细微差别的	R21
nuclear magnetic resonance 核磁共振	R22
number crunching 数字运算	R10

O

object [ˈɒbdʒɪkt] n. 物体 v. 不同意,不赞成,反对	4
obscure [əbˈskjʊə(r)] v. 使晦涩,使费解,使难懂	R4
obstacle [ˈɒbstəkl] n. 障碍,阻碍,绊脚石,障碍栅栏	9
off-duty 非值勤的,歇班的	13
offset [ˈɒfset] v. 使偏离直线方向 n. 偏离量,偏离距离	25
offshore platform 海洋平台	R13

onboard battery 机载电池	4
on-demand 按需的	2
on-premises 本地部署的（系统）	11
ontology [ɒnˈtɒlədʒi] n. 本体，本体论	R2
operational satellite 运行卫星	R11
optic [ˈɒptɪk] adj. 光学的，视觉的，眼睛的	22
optical character recognition（OCR）system 光学字符识别系统	R10
optical disc 光盘	R10
optical lithography 光学光刻	R18
optical proximity correction 光学邻近校正	R18
optimization [ˌɒptɪmaɪˈzeɪʃn] n. 最优化，充分利用	11
optimize [ˈɒptɪmaɪz] v. 优化，使最优化	3
orbit [ˈɔːbɪt] n.（天体等运行的）轨道，影响范围，势力范围	R11
orbital plane 轨道平面	R11
order tracking 订单跟踪	R16
orient towards 朝向，面对，确定方向，使适应	R3
original [əˈrɪdʒənl] adj. 起初的，原来的	10
outline [ˈaʊtlaɪn] v. 显示……的轮廓，勾勒……的外形，概述，略述	R4
overflow chute 溢流槽	R17
overhead crane 桥式起重机	R7
overheat [əʊvəˈhiːt] vi. 过热，愤怒起来 vt. 使过热，使愤怒 n. 过热，激烈	22
overlay [əʊvəˈleɪ] n. 覆盖，涂层 v. 覆在……上面，覆盖，铺	21
override [ˌəʊvəˈraɪd] n. 超控装置，预算超量，增加	16
overrun [ˈəʊvərʌn] n. 超出的成本（费用），超出的时间	R15
overturn [ˌəʊvəˈtɜːn] v. 倾倒，倾覆，翻掉	R5
oxide [ˈɒksaɪd] n. 氧化物	22
oxygen [ˈɒksɪdʒ(ə)n] n. 氧气	22

P

package [ˈpækɪdʒ] v. 将……包装好，包装成	3
packet [ˈpækɪt] n. 小包裹，（商品的）小包装纸袋，小硬纸板盒	R12
palletize [ˈpælətaɪz] v. 码垛堆积	6
palletizing robot 码垛机器人	R8
panel [ˈpænl] n. 面板	10
paradigm [ˈpærədaɪm] n. 样式，典范，范例	2
parallel [ˈpærəlel] adj. 并行的，平行的	1

parallel manipulator 并联机械手	6
parallelogram [ˌpærəˈleləɡræm] n. 平行四边形	6
parallelogram linkage system 平行四边形连杆系统	6
parameter [pəˈræmɪtə(r)] n. 参数	3
Pareto diagrams 帕累托图	R15
Pareto's law 帕累托法则	R15
part detection sensor 零件检测传感器	30
partially [ˈpɑːʃəli] adv. 部分地,不完全地	1
particle [ˈpɑːtɪkl] n. 微粒,粒子,颗粒	R9
partner [ˈpɑːtnə(r)] n. (合伙企业的)合伙人,(生意、组织或国家的)伙伴	26
passive [ˈpæsɪv] adj. 无源的,被动的	4
passive tag 无源标签	4
pattern recognition 模式识别	R10
payload [ˈpeɪləʊd] n. 净负荷,负载	R8
payload range 载荷范围	R8
pendant [ˈpendənt] n. 器,吊坠,(项链上的)垂饰	10
penetration [ˌpenəˈtreɪʃ(ə)n] n. 穿透,渗透	22
perceive [pəˈsiːv] vt. 察觉,注意到,认为,理解	R24
percentage [pəˈsentɪdʒ] n. 百分率,百分比	15
performance indicator 绩效指标	17
perimeter [pəˈrɪmɪtə(r)] n. 外缘,边缘	R5
periodically [ˌpɪərɪˈɒdɪkəli] adv. 定期地,周期性地	4
peripheral [pəˈrɪfərəl] adj. 外围的,周边的	15
peripheral material 外围材料	22
peristaltic [ˌperɪˈstæltɪk] adj. 蠕动的,蠕动引起的	9
peristaltic robot 蠕动机器人	9
pertinent [ˈpɜːtɪnənt] adj. 有关的,恰当的,相宜的	R4
pervasive [pəˈveɪsɪv] adj. 遍布的,充斥各处的,弥漫的	R1
petroleum [pəˈtrəʊliəm] n. 石油,原油	R8
pharmaceutical [ˌfɑːməˈsuːtɪkl] n. 药物 adj. 制药的	4
pharmaceutical plant 制药厂	R13
phase shift mask 相移掩模	R18
philosophy [fəˈlɒsəfi] n. 哲学	R1
photolithography [ˌfəʊtəlɪˈθɒɡrəfi] n. 光刻	R18
photo sculpture 照相雕刻法	24

photomask fabrication 光掩模制造	R18
physical [ˈfɪzɪk(ə)l] adj. 物理的,身体的,物质的	18
physical dimension 物理尺寸	18
pick and place 拾取与放置,贴装,贴片,取放	6
pitch axis 俯仰轴	15
pivotal [ˈpɪvətl] adj. 关键性的,核心的	R9
pixel [ˈpɪksl] n. 像素	R4
plane [pleɪn] n. 平面	25
plasma [ˈplæzmə] n. 等离子体	21
plasma control 等离子体控制	22
plastic [ˈplæstɪk] adj. 塑料的,人造的,不自然的	R8
plastic substrate 塑料基板	21
platform [ˈplætfɔːm] n. 平台,月台,讲台	R12
plug [plʌg] n. (电)插头,塞子 v. 堵,塞,补足	7
pneumatic [njuːˈmætɪk] adj. 气动的,压缩空气推动(操作)的,风动的	15
polarity [pəˈlærəti] n. 极性,截然对立,两极化	14
polarization voltage 极化电压	21
polish [ˈpɒlɪʃ] v. 擦光,磨光	6
pollution [pəˈluːʃ(ə)n] n. 污染,污染物	27
portable [ˈpɔːtəbl] adj. 便携式的,轻便的	14
portable terminal 便携式终端	15
porter [ˈpɔːtə(r)] n. 搬运工人,脚夫	R8
portion [ˈpɔːʃ(ə)n] n. 一份,一部分 v. 分配	7
positive direction 正向	15
potential [pəˈtenʃ(ə)l] n. 潜力,潜能,可能性 adj. 潜在的,可能的	5
powder [ˈpaʊdə] n. 粉,粉末	22
power [ˈpaʊə(r)] v. 驱动	4
precaution [prɪˈkɔːʃ(ə)n] n. 预防,警惕,预防措施	22
precision [prɪˈsɪʒn] n. 精确,准确,细致	5
precision casting 精密铸件	19
precision component 精密零件	R19
predefined [ˌpriːdɪˈfaɪnd] adj. 预定义的	24
predict [prɪˈdɪkt] v. 预测,预言	3
predictably [prɪˈdɪktəbli] adv. 可预言地,可预测地,可预料地	2
predictive analytics 预测分析	R21

predictive maintenance 预见性维修	3
prerequisite [ˌpriːˈrekwəzɪt] n. 先决条件，前提，必备条件	30
pressure [ˈpreʃə(r)] n. 压力 v. 对……施加压力	23
preventive [prɪˈventɪv] adj. 预防性的	3
preventive maintenance 预防性维护	3
previous [ˈpriːviəs] adj. 以前的，先前的，以往的	R12
prioritize [praɪˈɒrətaɪz] v. 划分优先顺序，优先处理，按重要性排列	11
prismatic [prɪzˈmætɪk] adj. 棱柱的，棱镜的	6
prismatic joint 移动关节	6
private cellular network 专用蜂窝网络	11
private network 专用网络	11
private sector organization 私营部门组织	27
proactively [ˌprəʊˈæktɪvli] adv. 积极主动地，主动出击地，先发制人地	2
probability [ˌprɒbəˈbɪləti] n. 概率，可能性	R1
procedural [prəˈsiːdʒərəl] adj. 程序上的，程序性的	R2
procedure [prəˈsiːdʒə(r)] n. 手续，步骤	19
process diagnosis 过程诊断	3
process planning 工艺规划	29
processing cell 加工单元	29
procurement [prəˈkjʊəmənt] n. 采购，（尤指为政府或机构）购买	R16
product life cycle management 产品生命周期管理	R13
product mix 产品结构	18
production [prəˈdʌkʃən] n. 生产，加工，产量	5
production automation 生产自动化	13
production floor 生产车间	13
production line 生产线	5
production log 生产日志	17
production scheduling 生产调度	1
production throughput 生产吞吐量	R18
productivity [ˌprɒdʌkˈtɪvəti] n. 生产率，生产力	R8
professional [prəˈfeʃənl] adj. 专业的，职业的	27
prognostics and health management 故障预测与健康管理	R23
programmable [ˈprəʊɡræməbl] adj. 可以编程的，计算机程序控制的	6
programmable logic controller 可编程控制器	20
programming [ˈprəʊɡræmɪŋ] n. （计算机）程序设计，程序编制，编程	R12

prohibitive [prəˈhɪbətɪv] adj. 令人望而却步的，禁止的，贵得买不起的	R14
project overrun 项目超支	R15
projection [prəˈdʒekʃn] n. 投射，投影	R22
prolific [prəˈlɪfɪk] adj. 多产的，创作丰富的	R1
prominent [ˈprɒmɪnənt] adj. 突出的，显眼的，显著的，凸显的	30
prompt [prɒmpt] n. 提示，提示符	20
propagation [ˌprɒpəˈɡeɪʃ(ə)n] n. 传播，扩展，宣传	R11
proportional [prəˈpɔːʃənl] adj. 相称的，成比例的	R10
prospect [ˈprɒspekt] n. 前景，可能性，希望，预期，展望	R23
protective [prəˈtektɪv] adj. 防护的，保护的	30
protective barrier 防护屏障	30
protocol [ˈprəʊtəkɒl] n. (数据传递的)协议，规程，规约	R1
prototyping [ˈprəʊtəʊtaɪpɪŋ] n. 原型设计	24
prototyping technology 原型技术	R17
provision [prəˈvɪʒn] n. 提供，供给，给养，供应品	R3
proximity [prɒkˈsɪməti] n. (时间或空间)接近，邻近，靠近	30
pseudo [ˈsjuːdəʊ] adj. 假的，冒充的	R11
psychology [saɪˈkɒlədʒi] n. 心理学	R1
pulling [ˈpʊlɪŋ] n. 拉销	18
pulp [pʌlp] n. 纸浆，浆状物，髓	14
pulsate [pʌlˈseɪt] v. 脉动，搏动，跳动，波动，振动，颤动，抖动	23
pulse [pʌls] n. 脉冲	4
pulse signal 脉冲信号	19
pump [pʌmp] n. 泵，抽水机 v. 抽吸	23
purge [ˈpɜːrdʒ] v. 清洗，净化	7
purging unit 清洁装置	7
push-button 按钮	15

Q

QR code QR 码	9
quality assurance 质量保证	13
quality control 品质控制	13
quantity [ˈkwɒntəti] n. 量，数量，大量	R12
quantum [ˈkwɒntəm] n. 量子，量子论，额(特指定额、定量)	19
quantum optical lithography 量子光学光刻	R18

R

radiation [ˌreɪdiˈeɪʃn] n. 辐射	4
radically [ˈrædɪkli] adv. 根本上，彻底地	R23
radio [ˈreɪdiəʊ] n. 无线电传送，收音机，广播电台	4
radio access network 无线接入网	11
radio receiver 无线电接收机	4
radio transmitter 无线电发射机	4
radio transponder 无线电应答器	4
radioactive [ˌreɪdiəʊˈæktɪv] adj. 放射性的，有辐射的	9
radio frequency identification 射频识别	4
radiographic [ˌreɪdiəʊˈɡræfɪk] adj. 胶片照相术的，射线照相术的	19
radioscopy [ˌreɪdiˈɒskəpi] n. 射线检查法，X光透视，放射线透视	19
radius [ˈreɪdiəs] n. 半径	25
random [ˈrændəm] adj. 随机的，任意的，胡乱的	R15
random variation 随机变化	R15
rapid prototyping（RP）快速原型	24
raster [ˈræstər] n. 光栅	R17
rationally [ˈræʃnəli] adv. 理性地	27
raw material 原材料，原料	3
react [riˈækt] v. 反应，回应	3
reaction [rɪˈækʃ(ə)n] n. 反应	22
reactive motion planning 反应式运动规划	30
reader device 阅读器	4
reagent [rɪˈeɪdʒənt] n. 试剂	21
real-time feedback 实时反馈	17
real-time sensing 实时传感	12
real-time synchronization 实时同步	12
real-time tracking 实时跟踪	13
receiver [rɪˈsiːvə(r)] n. 无线电接收机	4
recognition [ˌrekəɡˈnɪʃn] n. 识别，承认，表彰	19
reconfigure [ˌriːkənˈfɪɡə(r)] v. 重新配置(计算机设备等)，重新设定(程序等)	9
rectilinear [ˌrektɪˈlɪniə(r)] adj. 直线运动的	6
rectilinear robot 直线机器人	6
recycle [ˌriːˈsaɪkl] v. 再次应用，重新使用，回收利用	R3
reduction gearbox 减速箱	23
redundancy [rɪˈdʌndənsi] n. 多余，累赘	R5

redundant backup 冗余备份	R11
redundant DOF robot 冗余自由度机器人	30
redundant [rɪˈdʌndənt] adj. 冗余的，被裁减的，不需要的	R11
re-engineering 流程再造	27
reflective [rɪˈflektɪv] adj. 反射的，反光的	14
reflective tape 反光带	14
refuel [ˌriːˈfjuːəl] v. 换料，加燃料，加油	20
registration [ˌredʒɪˈstreɪʃn] n. 登记，注册，挂号	17
regularly [ˈreɡjələli] adv. 定期地，有规律地，频繁地，经常地	7
regulate [ˈreɡjuleɪt] v. 调节，控制（速度、压力等）	14
relaxation [ˌriːlækˈseɪʃən] n. 松弛的，弛张的	23
release [rɪˈliːs] v. 释放，放走，松开	R7
remarkable [rɪˈmɑːkəbl] adj. 引人注目的，非凡的	27
remote [rɪˈməʊt] adj. 远程的，远程连接的，偏远的，偏僻的	4
remote sensing 遥感	R22
remotely [rɪˈməʊtli] adv. 远程地，微弱地	R9
removal [rɪˈmuːvəl] n. 移走，去掉，消除，清除	23
remove [rɪˈmuːv] v. 移开，除去，废除	8
renewable [rɪˈnjuːəbəl] adj. 可再生的	27
renewable energy system 可再生能源系统	27
repetitive [rɪˈpetətɪv] adj. 重复的	R15
replace [rɪˈpleɪs] v. 代替，取代	1
repulsion [rɪˈpʌlʃn] n. 排斥力，斥力	R5
reputation [ˌrepjuˈteɪʃn] n. 名誉，名声	26
rescue [ˈreskjuː] v. 解救，援（营、挽）救，救出	5
reset [ˌriːˈset] v. 复位，重置	20
resilience [rɪˈzɪliəns] n. 恢复力，弹力，适应力	R9
resist [rɪˈzɪst] n. 抗蚀剂	R18
resist substrate 抗蚀剂衬底	R18
resolution [ˌrezəˈluːʃ(ə)n] n. 分辨率	R17
resolution enhancement technology 分辨率增强技术	R18
resource [rɪˈsɔːs] n. 资源，自然资源	26
respond [rɪˈspɒnd] v. 响应，（口头或书面）回答，做出反应，回应	26
restoration [ˌrestəˈreɪʃn] n. 恢复，复位	24
resultant [rɪˈzʌltənt] n. 生成物	21

retailer [ˈriːteɪlə(r)] n. 零售商,零售店　　　　　　　　　　　　　　　　R9
retool [ˌriːˈtuːl] v. 重组,重新装配　　　　　　　　　　　　　　　　　27
retrieval [rɪˈtriːv(ə)l] n. 检索,恢复,挽回,找回,取回　　　　　　　　　17
retrieve [rɪˈtriːv] v. 找回,收回,检索　　　　　　　　　　　　　　　　18
reuse [ˌriːˈjuːz] v. 再次使用,重复使用　　　　　　　　　　　　　　　27
revenue [ˈrevənjuː] n. 收入,收益,财政收入,税收收入　　　　　　　　R16
review [rɪˈvjuː] vt, n. 审查　　　　　　　　　　　　　　　　　　　　R15
revoke [rɪˈvəʊk] v. 取消,废除,使无效　　　　　　　　　　　　　　　R2
revolution [ˌrevəˈluːʃn] n. 旋转,旋转一周,革命　　　　　　　　　　　8
revolution counter 转数计数器　　　　　　　　　　　　　　　　　　　8
rework [ˌriːˈwɜːk] v. 返工　　　　　　　　　　　　　　　　　　　　　R15
rigidity [rɪˈdʒɪdəti] n. 刚性,强直,严格　　　　　　　　　　　　　　　R5
rinse [rɪns] v. 冲洗　　　　　　　　　　　　　　　　　　　　　　　　21
risk of fire 火灾危险　　　　　　　　　　　　　　　　　　　　　　　　23
robot [ˈrəʊbɒt] n. 机器人,自动操作装置,机器般的人　　　　　　　　　5
robot base 机器人底座　　　　　　　　　　　　　　　　　　　　　　　7
robot palletizer 机器人码垛工　　　　　　　　　　　　　　　　　　　R8
robot sensory signal 机器人感知信号　　　　　　　　　　　　　　　　5
robustness [rəʊˈbʌstnəs] n. 稳健性,健壮性　　　　　　　　　　　　26
role [rəʊl] n. 角色,任务　　　　　　　　　　　　　　　　　　　　　　22
roll axis 滚动轴　　　　　　　　　　　　　　　　　　　　　　　　　　15
roller mechanism 辊机构　　　　　　　　　　　　　　　　　　　　　R17
rotary [ˈrəʊtəri] adj. 旋转的,绕轴转动的,转动的　　　　　　　　　　　6
rotary coder 旋转编码器　　　　　　　　　　　　　　　　　　　　　　19
rotating [rəʊˈteɪtɪŋ] adj. 旋转的,轮值的　　　　　　　　　　　　　　14
rotation [rəʊˈteɪʃn] n. 旋转,转动　　　　　　　　　　　　　　　　　25
rough cut capacity planning 粗略缩减容量规划　　　　　　　　　　　R14
route sheet 工艺图表　　　　　　　　　　　　　　　　　　　　　　　18
routing file 工艺路线文件　　　　　　　　　　　　　　　　　　　　　R14
rugged [ˈrʌɡɪd] adj. 结实的,耐用的　　　　　　　　　　　　　　　　R5
run [rʌn] v. 运行,运转,运作,跑步　　　　　　　　　　　　　　　　　15

S

sales & operations planning 销售和运营计划　　　　　　　　　　　　R14
sales order entry 销售订单输入　　　　　　　　　　　　　　　　　　R14
satellite [ˈsætəlaɪt] n. 人造卫星,卫星　　　　　　　　　　　　　　　R11

sawing [ˈsɔːɪŋ] n. 锯切	18
scalable [ˈskeɪləbl] adj. 可扩展的,可去鳞的,可称量的	16
scanning lithography 扫描光刻	R18
scanning probe lithography 扫描探针光刻	R18
scene [siːn] n. 事件,场面,事发地,现场	R23
scene fusion 场景融合	13
schedule [ˈʃedjuːl] n. 计划(表),进度表,时间表	R15
scheduling [ˈʃedʒʊəlɪŋ] v. 调度,制定时间表 n. 行程安排	1
schematic [skiːˈmætɪk] n. (尤指电子电路的)示意图	29
scrap [skræp] v. 废弃	R3
scrap return 回炉料	12
screw [skruː] n. 螺钉	R6
screwdriver [ˈskruːdraɪvə(r)] n. 螺丝刀	7
scroll [skrəʊl] v. 滚屏,滚动,使相纸卷合(或打开)那样移动	16
seal [siːl] n. 密封,密封状态,水封	7
security [sɪˈkjʊərəti] n. 安全,保护措施	R9
segment [ˈsegmənt, segˈment] n. 部分,份,片,段	R11
seize [siːz] v. 抓住,捉住,把握(机会等)	26
self-contained 自成体系的,独立的,自立的	R13
self-organizing 自组织	3
semantic web 语义网	R2
semi-autonomy 半自动,半自主	5
semiconductor device 半导体器件	R18
sensitive [ˈsensətɪv] adj. 敏感的,体贴的,体恤的	R16
sensor [ˈsensə(r)] n. 传感器,敏感元件,探测设备	2
sensor detection 传感器检测	3
sensory [ˈsensəri] adj. 感觉的,感官的	R23
sensory data 感官数据	12
separate [ˈsepəreɪt] adj. 单独的,分开的,不同的,不相关的	29
sequence [ˈsiːkwəns] n. 顺序,次序,一系列,一连串	14
sequential [sɪˈkwenʃl] adj. 顺序的,序列的,按次序的	30
serial [ˈsɪəriəl] adj. 串联的,串行的	6
serial communication 串行通信	19
serial manipulator 串联机械手	6
service personnel 维修人员	7

词条	页码
servo ['sɜːvəʊ] n. (机器的)伺服系统,随动系统	30
servo drive 伺服驱动器	30
servomotor ['sɜːvəʊ.məʊtə] n. 伺服电机	23
shadow ['ʃædəʊ] n. 影子,阴影 v. 被……阴影笼罩	12
shaft [ʃɑːft] n. 轴,杆,柄,箭	16
shaft motion 传动轴	16
shaped tool 成型刀具	23
shift [ʃɪft] n. 轮班,工作时间改变,转变 v. (使)移动,(使)转移	16
shipbuilding ['ʃɪpbɪldɪŋ] n. 造船,造船业	22
shop floor 车间	16
short circuit 电路短路	19
shortcut ['ʃɔːtkʌt] n. 捷径,快捷方式(图标)	R1
sight [saɪt] n. 视力范围,视力,看见,视野	4
significantly [sɪɡˈnɪfɪkəntli] adv. 显著地,明显地	2
silicon oxide 二氧化硅	21
silicon substrate 硅基板	21
silo ['saɪləʊ] n. 料仓,筒仓	20
silo information 料仓信息	20
simulation [ˌsɪmjuˈleɪʃn] n. 模仿,仿真	2
simulation technology 仿真技术	R23
simulation verification 仿真验证	12
simultaneously [ˌsɪmlˈteɪniəsli] adv. 同时地,同步地	11
single cabinet controller 单机柜控制器	R7
single-piece workflow 单件工作流	28
single-chip computer 单片机	19
single-column e-beam 单列电子束	R18
single-jet technology 单射流技术	R17
singularity [ˌsɪŋɡjuˈlærəti] n. 奇点,异常,奇特,奇怪	30
skeleton ['skelɪtən] n. 骨骼,骨架,提纲	5
sketch [sketʃ] n. 素描,草图	24
sleeve [sliːv] n. (机器的)套筒,套管,袖子	7
slide [slaɪd] v. (使)滑行,滑动	6
slot [slɒt] n. 窄缝,(名单、日程或节目表中的)位置,时间	14
smear [smɪə(r)] n. 污迹,污渍,污点	R4
socket ['sɒkɪt] n. 承窝,承槽,插孔	R5

soft lithography 软光刻	R18
solar panels extended 太阳能电池板延伸	R11
solidify [səˈlɪdɪfaɪ] v. 使凝固	24
solubility [ˌsɒljuˈbɪləti] n. 溶解度	R18
solution [səˈluːʃn] n. 解决方案,溶液	3
sonar [ˈsəʊnɑː(r)] n. 声呐,声波定位仪	R22
sophisticated [səˈfɪstɪkeɪtɪd] adj. 复杂巧妙的,先进的,精密的	R2
space exploration 太空探索,外层空间探索	5
spam filtering 垃圾邮件过滤	R1
spare part 备件	18
spark [spɑːk] n. 火花,火星,诱因	23
spark channel 电花通道	23
spark machining 电火花加工	23
specialized [ˈspeʃəlaɪzd] adj. 专门的,专业的	17
species [ˈspiːʃiːz] n. 种类,物种	R21
specification [ˌspesɪfɪˈkeɪʃ(ə)n] n. 规范,明确说明,详述	7
specify [ˈspesɪfaɪ] v. 明确指出,把……列入说明书	18
spectrum [ˈspektrəm] n. 频谱,声谱,波谱,光谱,谱,范围,幅度	11
spindle [ˈspɪnd(ə)l] n. 轴,细长的人或物	16
spindle speed 主轴转速	16
splint [splɪnt] n. 夹板,薄木条	R8
splint mechanical gripper 夹板式机械抓手	R8
spotwelding cabinet 点焊机柜	R7
spray [spreɪ] v. 喷,喷洒,向……喷洒	6
spring [sprɪŋ] n. 弹簧,发条,春天,泉水	7
stack [stæk] n. (使)放成整齐的一叠(一摞、一堆),许多,栈	13
stacking [ˈstækɪŋ] n. 堆垛 v. 堆叠,堆积	R8
stainless steel 不锈钢	22
stakeholder [ˈsteɪkhəʊldə(r)] n. 利益相关者,参与人,参与方	R16
state-of-the-art 使用最先进技术的,体现最高水平的	R3
statistical [stəˈtɪstɪk(ə)l] adj. 统计的,统计学的	R15
statistical sampling 统计抽样	R15
statistics [stəˈtɪstɪks] n. 统计数字(或资料),统计学	R1
status [ˈsteɪtəs] n. 状况,情形	R20
steel [stiːl] n. 钢,钢铁	22

steering ['stɪərɪŋ] n.(车辆等的)转向装置	14
stiffness [stɪfnəs] n.刚度,硬度	6
stimuli ['stɪmjʊlaɪ] n.促进因素,激励因素,刺激物	R9
stowage cargo 码垛货物	R8
straight-line rapid positioning 直线快速定位	25
strapping ['stræpɪŋ] n.捆扎,皮带	R8
streamline ['stri:mlaɪn] v.精简(组织、流程等)使效率更高,使……成为流线型	R19
strength [streŋθ] n.强度,浓度,力量	23
string [strɪŋ] n.线,一连串,一系列(事件)	8
strive [straɪv] v.努力,力争	27
strong solvent 强溶剂	7
subcategory ['sʌbˌkætəgəri] n.子范畴,亚类	R20
subcontract [ˌsʌb'kɒntrækt] v.转包,分包	3
subjective [səb'dʒektɪv] adj.主观的,个人的,自觉的	19
subsequent ['sʌbsɪkw(ə)nt] adj.后来的,随后的	22
subset ['sʌbset] n.子集	24
substantial [səb'stænʃl] adj.大量的,价值巨大的,重大的	R15
substrate ['sʌbstreɪt] n.基片	21
sub-system 子系统	6
subtraction [səb'trækʃn] n.减去,(数)减,减法	R23
successive [sək'sesɪv] adj.继承的,连续的	24
suction ['sʌkʃ(ə)n] n.吸盘,吸力,抽吸	R8
superimposition [ˌsu:pərˌɪmpə'zɪʃn] n.添上,[摄]叠印,重叠	R23
supervise ['su:pəvaɪz] v.监督,管理,指导,主管	30
supervision [ˌsu:pə'vɪʒ(ə)n] n.监督,管理	17
supplement ['sʌplɪmənt] v.增补,补充 n.增补物,增刊	13
supply chain management system 供应链管理系统	26
supply chain 供应链	2
surface ['sɜ:fɪs] n.表面,水面,地面,桌面,台面	R19
surface micromachining 表面微细加工	21
surgically ['sɜ:dʒɪkli] adv.如外科手术般地	R19
surveillance [sɜ:'veɪləns] n.监控,监视	11
survival [sə'vaɪv(ə)l] n.存活,幸存	26
sustainable [sə'steɪnəbl] adj.可持续的	27

swan [swɒn] v. 无目的地漫游, 闲逛　　　　　　　　　　　　　　　R21

switch [swɪtʃ] v. 打开, 改变, 转变　　　　　　　　　　　　　　　R7

symbiotic [ˌsɪmbaɪˈɒtɪk] adj. 共生的, 互利的　　　　　　　　　　12

symbiotic evolution 共生演进　　　　　　　　　　　　　　　　　12

symbol [ˈsɪmbl] n. 符号, 象征, 标志, 代表性的人（物）　　　　　　5

synchronization [ˌsɪŋkrənaɪˈzeɪʃn] n. 同步　　　　　　　　　　　12

synchronize [ˈsɪŋkrənaɪz] v. 使同步, 在时间上一致, 同步进行　　R16

synthesis [ˈsɪnθəsɪs] n. 合成, 综合, 综合体　　　　　　　　　　R22

T

tackle [ˈtækl] v. 解决, 应付, 处理　　　　　　　　　　　　　　R10

tactile [ˈtæktaɪl] adj. 触觉的, 有触觉的, 能触知的　　　　　　　　9

tag [tæg] n. 标签　vt. 给……加上标签, 把……称作　　　　　　4

tank [tæŋk] n.（液体、气体、储藏）容器　　　　　　　　　　　　23

target [ˈtɑːɡɪt] n. 目标, 指标　　　　　　　　　　　　　　　　　15

teach pendant 示教器　　　　　　　　　　　　　　　　　　　　　8

technical [ˈteknɪk(ə)l] adj. 工艺的, 技术的　　　　　　　　　　　25

technical performance 技术绩效　　　　　　　　　　　　　　　R15

technique [tekˈniːk] n. 工艺, 技巧　　　　　　　　　　　　　　29

teleoperation [ˌteliˈɒpəˈreɪʃn] n. 远程操作　　　　　　　　　　　5

telepresence [ˈteliprezns] n. 遥现, 远程在位（利用计算机模拟过程）　5

temperature [ˈtemprɪtʃə(r)] n. 温度　　　　　　　　　　　　　23

template [ˈtempleɪt] n. 模板　　　　　　　　　　　　　　　　R4

temporary [ˈtemprəri] adj. 临时的, 短暂的　　　　　　　　　　28

temporary inventory 临时库存　　　　　　　　　　　　　　　　28

term [tɜːm] n. 术语, 专有名词　　　　　　　　　　　　　　　　27

terminal [ˈtɜːmɪnl] n. 终端, 终点站, 航站楼　　　　　　　　　　15

terrestrial terminal 地面终端　　　　　　　　　　　　　　　　R11

textual [ˈtekstʃuəl] adj. 文本的, 篇章的　　　　　　　　　　　R10

thermal [ˈθɜːm(ə)l] adj. 热的, 热量的　　　　　　　　　　　　22

thermal conduction 热传导　　　　　　　　　　　　　　　　　22

thermal scanning probe lithography 热扫描探针光刻　　　　R18

thermochemical nanolithography 热化学纳米光刻　　　　　R18

thermoset [ˈθəːməuset] adj. 热固性的　　　　　　　　　　　　24

thermoset polymer 热固性聚合物　　　　　　　　　　　　　　24

throughput [ˈθruːpʊt] n. 吞吐量, 生产量, 接待人数　　　　　　R8

throughput rate 通量率	R18
tight [taɪt] *adj*. 装紧的,密集的,挤满的	6
tighten [ˈtaɪtn] *v*. (使)变紧,更加牢固	R5
tightly [ˈtaɪtli] *adv*. 紧紧地,牢固地,紧密地	4
time-to-market 上市时间	2
tiny [ˈtaɪni] *adj*. 极小的,微小的,微量的	4
token bus 令牌总线	R20
token ring 令牌环	R20
tolerance [ˈtɒlərəns] *n*. 容忍,忍受	30
toll [təʊl] *n*. 通行费 *v*. 收费	R9
tool offset 刀具偏置	25
tool path 工具路径	R19
topography [təˈpɒɡrəfi] *n*. 地形学	24
torque [tɔːk] *n*. (使机器等旋转的)转矩	R5
tow [təʊ] *v*. 牵引,拖,拉,拽	14
toxic [ˈtɒksɪk] *adj*. 有毒的,引起中毒的,卑鄙无耻的	9
trace [treɪs] *v*. 追踪,查出,追溯找到,发现	3
tracked mobile robot 履带式移动机器人	9
trailer [ˈtreɪlə(r)] *n*. 拖车,挂车	14
trajectory [trəˈdʒektəri] *n*. 轨迹	10
transaxial tomography 轴向体层成像	R22
transform [trænsˈfɔːm] *v*. 转换,使变形,使转化	19
transmission [trænzˈmɪʃn] *n*. 播送,发射,发送,传输	4
transmit [trænzˈmɪt] *v*. 输送,发射,传播,传染,使通过	14
transmitter [trænzˈmɪtə(r)] *n*. (尤指无线电或电视信号的)发射机	4
transparency [trænsˈpærənsi] *n*. 透明度,清晰度	2
transponder [trænˈspɒndə(r)] *n*. 应答器,转发器	4
trap [træp] *v*. 使受限制,把……困在	R7
triangulate [traɪˈæŋɡjuleɪt] *v*. 对……进行三角测量,由三角形组成的	14
trigger [ˈtrɪɡə(r)] *v*. 触发 *n*. 触发器	4
trilateration [ˌtraɪlætəˈreɪʃn] *n*. 三边测量	R11
troubleshooting [ˈtrʌblʃuːtɪŋ] *n*. 解决难题,处理重大问题	3
true defect pixel 真实缺陷像素	19
tumor [ˈtjuːmə(r)] *n*. 肿瘤,肿块	R22
turnaround time 周转时间	R17

turret ['tʌrət] n. 塔楼,角楼,炮塔　　14

U

ubiquitous [juːˈbɪkwɪtəs] adj. 似乎无所不在的,十分普遍的　　R1
ubiquitous network 泛在网　　R24
ultra-high definition 超高清　　11
ultrasensitive sensing 超灵敏传感　　R18
ultrasonic scanning 超声波扫描　　R22
ultraviolet (UV) light 紫外线　　21
ultraviolet [ˌʌltrəˈvaɪələt] adj. 紫外的　　24
unergonomic [ˌʌnɜːgəˈnɒmɪk] adj. 不适于提升工效的　　30
unify [ˈjuːnɪfaɪ] v. 统一,使成一体,使一元化　　R3
unobtrusive [ˌʌnəbˈtruːsɪv] adj. 不引人注目的,不张扬的,不招摇的　　R9
unsupervised [ˌʌnˈsuːpəvaɪzd] adj. 无人监督的,无人管理的　　R21
uplink [ˈʌplɪŋk] n. 上行链路　　11
upper bevel gear 上锥齿轮　　7
upstream [ˌʌpˈstriːm] adj. 向(在)上游的,逆流而上的　　R23
uptime [ˈʌptaɪm] n. 运行时间,上线时间　　17
user interface 用户界面　　R2
utility [juːˈtɪləti] n. 应用程序,实用程序　　16
utilization [ˌjuːtəlaɪˈzeɪʃn] n. 利用,使用　　27
utilize [ˈjuːtəlaɪz] v. 利用,使用,运用,应用　　11

V

vacuum [ˈvækjuːm] adj. 真空的 n. 真空容器　　R8
vacuum suction mechanical gripper 真空吸取式机械抓手　　R8
valid [ˈvælɪd] adj. 有效的　　3
value stream 价值流　　28
vapor [veɪpə] n. 蒸汽,烟雾　　22
vaporize [ˈveɪpəraɪz] v. (使)蒸发　　21
variable [ˈveərɪəb(ə)l] n. 变量,变数　　R15
variance [ˈveərɪəns] n. 变化幅度,分歧,不一致,方差　　19
variation [ˌveərɪˈeɪʃn] n. 变化,变动　　26
vaseline [ˈvæsəliːn] n. 凡士林　　7
vat [væt] n. 大桶　　24
vehicle [ˈviːəkl] n. 交通工具,车辆　　4
vehicle airbag accelerometer 车辆安全气囊加速计　　21

velocity [vəˈlɒsəti] n. (沿某一方向的)速度	2
vendor [ˈvendə(r)] n. 供应商,销售公司,摊贩,(房屋等的)卖主	11
verification [ˌverɪfɪˈkeɪʃ(ə)n] n. 验证,证明	12
vertical [ˈvɜːtɪk(ə)l] adj. 垂直的,直立的,纵向的	7
vertical arm 垂直臂	7
vertically [ˈvɜːtɪkəli] adv. 垂直地,直立地	6
viable [ˈvaɪəbl] adj. 可行的,可实施的,能独立发展的	R18
violent [ˈvaɪəl(ə)nt] adj. 暴力的,猛烈的	22
virtual [ˈvɜːtʃuəl] adj. 虚拟的,模拟的	2
virtual copy 虚拟副本	2
virtual instrument 虚拟仪器	R12
virtual real mapping 虚实映射	12
virtual reality fusion technology 虚拟现实融合技术	12
virtual reality 虚拟现实	3
virtual simulation 虚拟仿真	2
virtualized encapsulation 虚拟封装	R3
virtually [ˈvɜːtʃuəli] adv. 事实上	24
visible [ˈvɪzəbl] adj. 明显的,看得见的	22
visual [ˈvɪʒuəl] adj. 视觉的,视力的,可见的	4
visualization [ˌvɪʒuəlaɪˈzeɪʃn] n. 形象化,可视化	24
visualization technology 可视化技术	R8
void [vɔɪd] n. 空洞,空隙,空白 adj. 不合法的	19
void content 空洞率	19
volatile [ˈvɒlətaɪl] adj. 易变的,动荡不定的,反复无常的	R2
voltage [ˈvəʊltɪdʒ] n. 电压	23

W

walkthrough [ˈwɔːkθruː] n. 巡回检查	R15
warehouse [ˈweəhaʊs] n. 仓库,货栈,货仓	11
washer [ˈwɒʃə(r)] n. (螺母等的)垫圈,垫片,衬垫,洗衣机	R6
waste [weɪst] n. 废物,废弃物 v. 浪费,滥用,未能充分利用	27
water-cooled mirror 水冷反射镜	22
wattage [ˈwɒtɪdʒ] n. 瓦数	25
wattage system 瓦数系统	25
wear [weə(r)] v. 磨损,穿(衣服)	3
wearable computing 穿戴式计算	R9

weigh [weɪ] v. 重，称重	R11
weld [weld] v. 焊接，熔接，锻接	6
weld seam 焊缝	19
wet etching 湿蚀刻法	21
wheeled [wiːld] adj. 轮式的，有轮的	9
wheeled mobile robot 轮式移动机器人	9
wireless cellular connectivity 无线蜂窝连接	11
wireless sensor 无线传感器	R9
wisdom manufacturing 智慧制造	R24
workflow ['wɜːkfləʊ] n. 工作流程	3
workflow control system 工作流控制系统	26
working capital 营运资金	R14
workpiece ['wɜːkpiːs] n. 工件	3
workpiece center 工件中心	18
workpiece coordinate system 工件坐标系	25
workshop ['wɜːkʃɒp] n. 车间	3
wrist [rɪst] n. 手腕，腕关节	R6

X

X-ray imaging X 射线成像	19
X-ray lithography X 射线光刻	R18
x-y-z robot 三坐标机器人	6

Y

yield [jiːld] v. 产生，放弃，让步 n. 产量，收益，利润，红利	19

Z

zero downtime 零停机	28
zero inventory 零库存	28
zero latency 零时延	28

Appendix II Abbreviations

3GPP (3rd Generation Partnership Project)　　第三代合作伙伴计划
ADC (Analog-to-digital Converter)　　模数转换
AGV (Automatic Guided Vehicle)　　自动导引车
AI (Artificial Intelligence)　　人工智能
AIDC (Automatic Identification and Data Capture)　　自动识别和数据捕获
AM (Additive Manufacturing)　　增材制造
AM (Agile Manufacturing)　　敏捷制造
AM (Asset Monitor)　　资产监控
AMR (Autonomous Mobile Robot)　　自主移动机器人
AR (Augmented Reality)　　增强现实
ARAT (Active Reader Active Tag)　　主动阅读器主动标签
BGA (Ball Grid Array)　　球栅阵列封装
BOM (Bill of Material)　　物料清单
CAD (Computer-aided Design)　　计算机辅助设计
CAM (Computer-aided Manufacturing)　　计算机辅助制造
CAPP (Computer-aided Process Planning)　　计算机辅助工艺设计
CC (Cloud Computing)　　云计算
CCD (Charge Coupled Device)　　电荷耦合器件
CDMA (Code Division Multiple Access)　　码分多址
CIES (Centralized Integrated Enterprise System)　　集中式集成企业系统
CIM (Computer-integrated Manufacturing)　　计算机集成制造
CM (Cellular Manufacturing)　　单元式制造
CNC (Computer Numerically Control)　　计算机数控
CPS (Cyber-physical System)　　信息物理系统
CPU (Central Processing Unit)　　中央处理器
CRP (Capacity Requirement Planning)　　容量需求规划
CSR (Corporate Social Responsibility)　　企业社会责任
DAC (Digital-to-analog Converter)　　数模转换
DCS (Distributed Control System)　　分散控制系统
DDM (Direct Digital Manufacturing)　　直接数字化制造
DOF (Degree of Freedom)　　自由度
DRP (Distribution Resource Planning)　　分配资源计划
EBDW (Electron-beam Direct-write)　　电子束直写
EBL (Electron-beam Lithography)　　电子束光刻

EDM（Electrical Discharge Machining）	电火花加工
ERP（Enterprise Resource Planning）	企业资源计划
EUVL（Extreme Ultraviolet Lithography）	极紫外光刻
FCS（Finite Capacity Scheduling）	有限容量调度
FIFO（First Input First Output）	先进先出
FMS（Flexible Manufacturing System）	柔性制造系统
GPS（Global Positioning System）	全球定位系统
HAZ（Heat Affected Zone）	热影响区
HMI（Human Machine Interface）	人机界面
HRI（Human-robot Interaction）	人机交互
HSM（High-speed Machining）	高速加工
IATF（Information Assurance Technical Framework）	信息保障技术框架
IBD（Industrial Big Data）	工业大数据
IIoT（Industrial Internet of Things）	工业物联网
IM（Intelligent Manufacturing）	智能制造
IoP（Internet of People）	人联网
IoT（Internet of Things）	物联网
JIT（Just in Time）	准时制生产方式
LCM（Light Coupling Mask）	光耦合掩模
LGV（Laser Guided Vehicle）	激光导引车
LTE（Long Term Evolution）	长期演进
MAS（Multi-agent System）	多智能体系统
MBE（Model-based Enterprise）	基于模型的企业
MES（Manufacturing Execution System）	制造执行系统
MIS（Management Information System）	管理信息系统
ML（Machine Learning）	机器学习
MPS（Master Production Schedule）	主生产计划
MRP（Manufacturing Resource Planning）	制造资源计划
MRP（Material Requirement Planning）	物资需求计划
NGL（Next Generation Lithography）	下一代光刻
NL（Nanolithography）	纳米光刻
OCR（Optical Character Recognition）	光学字符识别
OEE（Overall Equipment Effectiveness）	设备综合效率
OPC（Optical Proximity Correction）	光学邻近校正
PCB（Printed Circuit Board）	印刷电路板
PCS（Process Control System）	过程控制系统
PDM（Product Data Management）	产品数据管理
PHM（Prognostics and Health Management）	故障预测与健康管理
PLC（Programmable Logic Controller）	可编程逻辑控制器

PRAT (Passive Reader Active Tag) 被动读卡器主动标签
PSM (Phase Shift Mask) 相移掩模
QC (Quality Control) 品质控制
QOL (Quantum Optical Lithography) 量子光学光刻
QR Code (Quick Response Code) 快速响应码
RAM (Random Access Memory) 随机存储器
RFID (Radio Frequency Identification) 射频识别
RP (Rapid Prototyping) 快速原型
RPA (Robotic Process Automation) 机器人流程自动化
RT (Radiographic Testing) 射线检验
SCADA (Supervisory Control and Data Acquisition) 监控与数据采集
SPL (Scanning Probe Lithography) 扫描探针光刻
SCM (Single-chip Computer) 单片机
SFC (Shop Floor Control) 车间控制
SMT (Surface Mounting Technology) 表面贴装技术
TP (Teach Pendant) 示教器
TPM (Total Productive Maintenance) 全面生产维护
TQM (Total Quality Management) 全面质量管理
VR (Virtual Reality) 虚拟现实
WIP (Work in Progress) 在制品

References

[1] 宋庭新,娄德元.先进制造技术(英文版)[M].武汉:华中科技大学出版社,2021.
[2] 范君艳,樊江玲.智能制造技术概论[M].2版.武汉:华中科技大学出版社,2022.
[3] 吴让大.高功率激光切割设备与工艺[M].武汉:湖北科学技术出版社,2010.
[4] 刘骋,蔡静,刘小芹.电子与通信技术专业英语[M].5版.北京:人民邮电出版社,2019.
[5] 刘小芹,王珏.现代制造技术专业英语[M].3版.武汉:华中科技大学出版社,2015.
[6] 刘小芹,张敬衡.数控技术专业英语[M].3版.北京:高等教育出版社,2015.
[7] MONK E,WAGNER B.Concepts in enterprise resource planning[M].2nd ed.Boston:Course Technology Press,2012.
[8] POCHET Y,WOLSEY L A.Production planning by mixed integer programming[M].New York:Springer,2006.